SENSES OF UPHEAVAL

SENSES OF UPHEAVAL

PHILOSOPHICAL SNAPSHOTS OF A DECADE

MICHAEL MARDER

ANTHEM PRESS

Anthem Press
An imprint of Wimbledon Publishing Company
www.anthempress.com

This edition first published in UK and USA 2022
by ANTHEM PRESS
75–76 Blackfriars Road, London SE1 8HA, UK
or PO Box 9779, London SW19 7ZG, UK
and
244 Madison Ave #116, New York, NY 10016, USA

Copyright © Michael Marder 2022

The author asserts the moral right to be identified as the author of this work.
All rights reserved. Without limiting the rights under copyright reserved above,
no part of this publication may be reproduced, stored or introduced into
a retrieval system, or transmitted, in any form or by any means
(electronic, mechanical, photocopying, recording or otherwise),
without the prior written permission of both the copyright
owner and the above publisher of this book.

British Library Cataloguing-in-Publication Data
A catalogue record for this book is available from the British Library.

Library of Congress Control Number: 2021949720

ISBN-13: 978-1-83998-226-2 (Hbk)
ISBN-10: 1-83998-226-8 (Hbk)
ISBN-13: 978-1-83998-229-3 (Pbk)
ISBN-10: 1-83998-229-2 (Pbk)

Cover credit: Drift, Mid-Atlantic Ridge, Anaïs Tondeur 2014.

This title is also available as an e-book.

For Slavoj, from whom I've learned so much, in friendship

CONTENTS

A Sense of Upheaval   1

PART I. POLITICAL UPHEAVAL

1. Rating Sovereignty   7
2. The Unfinished Collapse of the Soviet Union   11
3. We, the Orphans of October   15
4. Incendiary Words and the Volcano of Occupation   19
5. Can There Be Poetry after Netanyahu?   23
6. Marginalizing Europe   25
7. The European Union and the Rhetoric of Immaturity   29
8. Trump Metaphysics   33
9. The Con Artistry of the Deal: Trump, the Reality TV President   37
10. Covid-19: This Is Not a War   41
11. Going Viral, or The Coronavirus Is Us   45
12. Can Democracy Save the Planet?   49

PART II. CULTURAL UPHEAVAL

1. On Knees and Elbows   55
2. Being in Exile from Oneself   59
3. The Muslim "No"   61
4. Don't Keep Calm! And Don't Carry On!   65
5. Uncultured Austerity   69

6. A Genealogy of Enjoyment   71
7. The Two Suns of Europe   75
8. For the Love of a City   79
9. What Horse Meat Tells Us about Ourselves   81
10. Contagion: Before and after Covid-19   83

PART III. INTELLECTUAL UPHEAVAL

1. A Fight for the Right to Read Heidegger   89
2. Heidegger's Thinking Today Is, Perhaps, the Possibility of the World 93
3. *Plus de restes*: Remembering Jacques Derrida   97
4. The Philosopher's Beard   101
5. Naturalize This! Analytic Philosophy and the Logic of Reactive
   Neutralization   105
6. Jokes and Their Relation to Crisis   111
7. Position as a Political Category: Phenomenology and the Eroticism
   of Power   115
8. The Powerlessness of Philosophy   121

PART IV. TECHNOLOGICAL UPHEAVAL

1. Chernobyl as an Event   127
2. Nuclear Mourning   131
3. The Meaning of "Clean Energy"   133
4. Without Clean Air, We Have Nothing *(with Luce Irigaray)*   137
5. Poland's Bialowieza: Losing the Forest and the Trees   141
6. Just Randomness?   145
7. The Idea of Following in the Age of Twitter   151

The Upheavals Yet to Come   155

Notes   157

A SENSE OF UPHEAVAL

We live in a time of great upheavals. Every sphere of existence is now home to the destabilizing forces that are drastically changing the environment, redrawing social boundaries, shaking up the economy and family relations, altering individual psychologies and political orders. Catastrophe, global devastation, and apocalypse no longer refer to fantastic scenarios waiting to unfold in a dystopian future; they knock on our doors. How to interpret the world in a state of upheaval, a world that, more swiftly and more deftly than ever before, seems to elude our capacities for understanding reality?

Karl Marx famously wrote in his eleventh thesis on Feuerbach, "The philosophers have only *interpreted* the world in various ways; the point, however, is to *change* it."[1] A rejoinder to Marx may be that the point now, in the twenty-first century, is to interpret a changing world and that, in any case, without a careful and indefatigable interpretation, it is impossible to ascertain *if* and *how* the world is changing. My contention in this book is that the world is changing through upheavals that, when experienced from within, cause it to lose its quality of worldliness, its livable, habitable character.

We treat upheavals as more or less synonymous with revolutions, mayhem, disorder, disarray. Still, there is something highly idiomatic about this word and the process it designates, something that will help shed light (however unsteady) on what is going on at present.

Upheaval literally means the heaving up of the earth. The English verb *to heave* derives from the Proto-Germanic **hafjan* and entails the raising of something, lifting up what used to be below. It is closely related to the more common verb *to have*, to take hold of, to possess. In upheavals, though, taking hold slips away from the grasp and control of those who experience—or undergo without being able to form a full-fledged experience—extreme events. It is not we who take hold of the tendencies underway; rather, these tendencies grip us, throwing us, along with our world, up in the air. We are deprived of the support

1

we usually seek from the more or less stable foundations of our existence in the social and cultural meshwork of customs, in a habitual relation to our natural environment, in the daily routines of work or leisure. The dynamics of upheaval are close to elemental forces, especially those related to the element of the earth with its seismic activity, entirely uncontrollable and only barely predictable.

As in the churning up of the earth, what is lifted in upheaval is what was already there as the obverse of the present: secret, invisible, obscure, abstruse... With the shadowy underside of existence—of our own existential, social, economic, political, or ecological being—emerging in the daylight of the present, a certain version of the apocalypse, which implies the ultimate uncovering or unveiling of reality, comes to fruition. That said, apocalyptic upheavals are not just drawing the curtain that has prevented us from seeing the true nature of reality; they meddle with the system of coordinates for meaningful experience, upending things, throwing what was down up and bringing what was up down. More radical yet than a revolution, a persistent upheaval invalidates our tried-and-tested methods for orienting ourselves in space, time, and the pluriverse of sense. I have dealt with this totally disorientating aspect of upheaval in my *Dump Philosophy: A Phenomenology of Devastation*, as well as in *Pyropolitics in the World Ablaze*.[2]

Not by chance, upheaval is a kindred word of the German *Aufhebung*, which it translates into English with much more precision than "sublation," "reversal," "removal," or "revocation." This word, this concept, is one of the keys to Hegel's dialectics, where it plays the role of the engine for the development of the most disparate entities, from plants to the institutions of the state, from human consciousness to artworks, from logical categories to organic systems. While *Aufhebung* concentrates in itself the labor of the negative (i.e., of negating the preceding state of whatever or whoever is subject to it), it is indispensable to the positive development of the entity in question. The same goes for upheaval, which, besides its destructive or disorienting connotations, renders possible what was impossible before. More precisely, upheaval allows for the disclosure of the murky layers of being or nonbeing, silently subtending things, not to mention the switch of positions that sets an otherwise static order in motion. Perhaps we should say with Hegel that everything and everyone develops by upheaval, not in the comfort of an undisturbed self-identity but thanks to the splitting open of this identity, its churning and reversal, the uplifting of what was suppressed or repressed.

With regard to the negative and positive poles of upheaval, this book casts a panoramic glance at the decade of crises, stretching between the years 2010 and 2020. Although it reunites my often very targeted and time-sensitive interventions in public debates, the text exceeds the contextual frames wherein it was originally embedded and paints a picture of upheavals in four broad areas: politics, culture, intellectual life, and technology. When necessary, each chapter is preceded by a brief summary of the circumstances in which it initially arose. The arrangement of texts here is not chronological but thematic, ensuring that issues well in excess of their unique historical situation come to the fore with utmost clarity, crispness, and precision. In short, this book is an experiment in interpreting a changing world, the world of upheavals and in upheaval.

PART I

POLITICAL UPHEAVAL

1
RATING SOVEREIGNTY

In the financial crisis that marked the beginning of the decade, rating agencies, including Fitch or Standard & Poor (S&P), lowered the credit ratings of many Southern European states to the level of "junk." This development further impoverished the already fragile national economies unable to issue state bonds with positive interest in order to finance portions of their budgets. It drove the politics of austerity in the European Union and prompted a rollercoaster ride of financial speculation with sovereign debts.

One of the most significant political lessons of the Eurozone crisis is that the classical model of sovereignty has ceased to work. As the examples of Greece, Portugal, Italy, and Spain have shown, the nation-state has lost the last shreds of its supreme authority to make decisions on domestic policy matters.

Debtor states and candidates to the infamous "bankruptcy club" of countries likely to declare economic default must play by the rules formulated outside of their jurisdiction if they are to ensure borrowing rates conducive to their continued financing and, indeed, survival. In return, external political and extra-political bodies such as the Troika—the European Union, European Central Bank, and International Monetary Fund—gain the right to dictate the age of retirement, minimum wages, the length of the workweek, or the extent of salary and pension cuts, among other policies to be adopted by affected EU member states.

Having turned into mere appendages of foreign financial and political agencies, the national governments of Portugal and Greece are reduced to the superfluous role of signing off on their decisions. The vocabulary of an antiquated paradigm of sovereignty and national self-determination is still used, even as its terms and coordinates have grown obsolete. In Europe, a distinct, albeit yet unnamed, model of sovereignty is rapidly being forged.

Behind the scenes, rating agencies, such as S&P, have had a hand in the demise of sovereign states. As countries increasingly finance themselves through market borrowing, their solvability and the sustainability of their debt levels provide the fodder for intense financial speculation. This is where rating agencies come in: by lowering the credit ratings of sovereign debts, they push borrowing costs of the rated states up, forcing them, in the last resort, to seek assistance from—and to deliver themselves to the whims of—the Troika.

Formal national sovereignty is of little consequence when faced with the oracles of the rating agencies that have the power to reduce sovereign credit status to "junk." Needless to say, the judgments of the emerging decision-makers are less than reliable. A recent court case in Australia has demonstrated that S&P misled investors about the soundness of their investments by giving the highest AAA rating to the riskiest of securities. Why should we be surprised? After all, this is consistent with the upside-down worldview that attributes the lowest ratings to sovereign debts guaranteed by European countries that are much more stable than derivatives and hedge funds.

As a philosopher, I am interested in the framework of knowledge production and organization, which amounts to the "regime of truth" underpinning rating activity. Rating agencies are only a sign of the times; rankings are prevalent in all areas of life, from academic journals and universities to restaurants. The advantage is that consulting a ranking helps us make fast decisions on where best to dine or publish a research paper. But this presupposes that the end results, whereby ranked materials are sequentially ordered from the highest to the lowest, merely organize the information that is already there, without either adding to or subtracting from it, let alone questioning the criteria according to which rankings operate.

Problems crop up when, thanks to inconsistencies, vague criteria, or predetermined expectations (also known as biases), the ranking creates a world of its own, which has little to do with the situation at hand. This is, precisely, what happened in the case of credit rating agencies that, instead of reflecting the state of the world, have conjured one up. At odds with the actual level of risk, this constructed world crumbled as soon as a crisis revealed the real extent of financial speculation and burst the bubble of fake security vouchsafed by *triple-A* grades.

The next question we might raise is, "What kind of a world is it, where everything is eminently rankable?" The answer is obvious: it is a world in which everything has been qualitatively flattened and where (ideally) only quantitative differences persist. What rankings organize is information, rather than knowledge, and that is why they are so appropriate to today's Information Age.

In this, they are consistent with the modern scientific approach that strives to translate all of material reality into numbers, equations, and algorithms, paving the way to a total quantification and "operationalization" of the world.

The issue is not only that, as in the case of rating agencies, numbers and numeric values do not match the state of affairs they purportedly transcribe, but also that qualitative differences are largely lost even in the most accurate process of the world's mathematization. While the hierarchies of valuation are probably as old as humanity itself, the numerical (and, hence, pseudoscientific) parceling of values is a relatively new phenomenon, which threatens the very fabric of ethical and political existence.

When the dogma of shoddily constructed rankings presents itself in the guise of truth, and when we look to its products to provide us with efficient shortcuts to informed decisions, knowledge societies deteriorate to the status of ignorance societies. And, then, it takes nothing less than a full-blown crisis to jolt us back to our senses.

There is, however, at least one exception to the general rule of flattening and to the lack of qualitative differentiation characteristic of modernity and its offshoot, the obsession with ratings. This exception is sovereignty.

Sovereignty has been traditionally understood in terms of the highest political clout, a power that is supreme and absolute. Following this precept, relations among sovereigns are not subject to any other authority, which is why, for Thomas Hobbes, international relations (i.e., relations among sovereign states) are hopelessly mired in the state of nature and overshadowed by the perpetual possibility of war.

We might say, in turn, that sovereignties are unrankable and unratable, as they are all equally supreme and absolute within their respective domains. They are simply incomparable, much less available for external adjudication by a third party that would, with reference to its decision-making authority, trump their own supremacy.

Modernity has always been mistrustful of sovereignty, in which it—perhaps rightly—sees the anachronistic vestiges of premodern absolutism. The division of powers into executive, judiciary, and legislative branches in parliamentary democracy is a not-so-concealed attempt to do away with the concept of sovereignty altogether by imposing a system of checks and balances on any authority that may claim for itself the title "supreme."

The current disintegration of sovereignty is the culmination of a long history that saw this concept and the corresponding institution gradually eroded. Still, if sovereignty has been resilient thus far, it is because decision-making has proven, in the last instance, indispensable. This means that sovereignty is not really destroyed but, rather, displaced, relocated elsewhere, as it was in its historical transfer from the body of the king to the ideal body of the people in the aftermath of the French Revolution.

The paradox and the novelty of the displacement we are witnessing today is that the decision-making bodies that have appropriated whatever remains of sovereignty have done so in a selective manner. The sovereignty unofficially invested in the Troika and, even less overtly, in international rating agencies is all about imposing decisions without assuming any political, legal, or financial responsibility for them, let alone without legitimately representing a body of population.

Apparently emanating from nowhere—arising *ex nihilo*—the effects of sovereign verdicts are clothed in a mantle of inevitability and inscrutability. We are led to believe that arguing against them amounts to going against the iron law of the market, heedless of any alternatives and, for that matter, devoid of freedom. But it is not enough to protest against these most recent embodiments of fate that hold the destinies of entire countries in their hands. To begin with, we must consider the real root of the problem: the exceptional meaning of sovereignty in an increasingly flattened, quantitatively defined, and rankable world.

2

THE UNFINISHED COLLAPSE
OF THE SOVIET UNION

In 2020, many of the former Soviet republics were beset with wars and tumult: from the armed conflict between Armenia and Azerbaijan in the Nagorno-Karabakh region to the ongoing mass protests against official election results in Belarus.

Many international commentators are gleefully considering the war and unrest now engulfing Belarus, Eastern Ukraine, Armenia, Azerbaijan, and Kyrgyzstan as nothing more than challenges to Russia's (and, personally, Vladimir Putin's) sphere of influence in "post-Soviet space." *Financial Times* declares "Russia's neighborhood" to be "in flames";[1] *Bloomberg News* gloats that "it's harder and harder for Vladimir Putin to call the shots" in Russia's "near abroad";[2] *The New York Times* proclaims that "Putin, long the sower of instability, is now surrounded by it."[3]

What these and many other related analyses have in common is their historical shortsightedness, which has come to dominate our epoch of clickbait news, sound bites, and on-demand mass production of opinions. It is simply assumed that the collapse of the USSR is an event that happened in the past, a roughly three-year process of dissolution that culminated in replacing the Soviet red flag with the Russian tricolor over the Kremlin on December 25, 1991. But are things as simple as that? A political entity that, in a world-historical upheaval, was born from the first successful workers' revolution, that was crucial to the defeat of Nazism in World War II, and that was engaged in a protracted Cold War with the other superpower, the United States: does the disappearance of such a political actor limit itself to a meager three-year period?

To put the strife in which some of the former Soviet republics are engulfed in its proper context, we need to develop a robust philosophy of history, above all of political history. I do not mean that what is necessary is a hairsplitting production of academic articles and books, reveling in jargon that remains inaccessible to the common folk. Instead, at the level of our public consciousness, now shaped by the rather one-dimensional commentators working for mainstream news organizations, we need to imagine political history otherwise, with its gaps, protracted subterranean processes, and time lags between causes and effects.

There is a well-known anecdote about the conversation US secretary of state Henry Kissinger had with Chinese Prime Minister Zhou Enlai in the early 1970s. When asked what he thought of the French Revolution of 1789, the Chinese statesman said, "It is too early to tell." In an attempt to explain away this puzzling response to an event that had taken place nearly two hundred years prior to the conversation, it was said that, in a translation glitch, Zhou Enlai thought he was being asked about the 1968 student uprisings in Paris. To Western commentators it was simply inconceivable that something that had happened so long ago was still not open to historical judgment and could still have an indeterminate future. But, even if some sort of a confusion between French events in the eighteenth and the twentieth centuries took place, it was wholly justified: were not the uprisings of 1968 the afterglows of the French Revolution, alongside the Russian, Chinese, Cuban, and other revolutions striving for universal human emancipation through the emancipation of the workers?

And so, to us, living less than thirty years after the "official" collapse of the Soviet Union, it should be obvious that it is still too early to tell what this event has meant and whether it has even been completed. In the East and in the West, our lives are invisibly molded by its aftermath, as though our recent histories are a comet tail of that momentous occurrence.

The ongoing dismantlement of the welfare state in Canada, the United States, and Europe is largely explicable with respect to the disappearance of the Soviet adversary and a related radical shift in the global ideological struggle, which previously required a quasi-socialist placating of workers in the West. The short-lived assertion of a hegemonic world superpower status by the United States, reflected in Francis Fukuyama's narrative of the neoliberal end of history, was equally an outcome of that collapse, its consequences still far from exhausted. Within Russia itself, an intergenerational, cultural, and even communicational or linguistic gap marks a visible scar on the body politic. And, of course, in Belarus, Ukraine, Armenia, Azerbaijan, and Kyrgyzstan, various contestations of the status quo are symptoms of the non-resolved legacies of the USSR's decline.

It is against this background that our relation to big historical events needs to be carefully recalibrated. Once again, cosmic analogies come to mind: it takes solar energy, traveling at the speed of light, 8 minutes and 20 seconds to reach the earth; similarly, there is a delay between a significant historical (or world-historical) event and its "arrival," the full unfolding of its consequences. With reference to the breakdown of the Soviet Union, our existence is still happening in this delay, just like a ray of the sun that, having left the solar corona, has not yet hit the surface of the earth.

This means two things. First, we know that we can make sense of history only in retrospect, but the exact point in time when such making-sense can happen with any degree of certainty is, itself, unclear. Unlike in astrophysics, where calculations allow us to predict when the energy released by the sun would make it to our planet, there is no such technique in the field of historical studies. Second, the indeterminacy of an event and of its many consequences implies that it is radically open, unstable even. It follows that, still on its way, an event may come to the fullness of its accomplishment *or* be counteracted and derailed: the collapse of the USSR may, itself, collapse, or be brought to completion. In this sense, possibility, with the indeterminacy it signals, derives from a living past, not from the future.

3

WE, THE ORPHANS OF OCTOBER

The year 2017 was the centennial anniversary of the Russian revolution. This essay was written as a commemoration of that event.

It's November 7, 1987. Along with hundreds, if not thousands, of other children I am in the main hall of Moscow's Revolution Museum, since renamed State Central Museum of the Contemporary History of Russia. On its seventieth anniversary, the Revolution of 1917 had been already thoroughly museumized, even as Mikhail Gorbachev was initiating a series of reforms intended to revive its legacy and shake off decades of bureaucratic torpor. His *perestroika*, translatable as *reconstruction*, envisioned a re-revolutionizing of society and a renewal of the impulse behind the Red October.

At the museum, we were awaiting the beginning of the ceremony marking our induction into *Oktyabryata*, literally *small October'ites*, the children of October. That was the first step on the ladder leading to full-fledged membership in the Communist Party and passing from *Oktyabryata* through *Pioneers* to *Communist Soviet Youths* (*Komsomol*). With plenty of excitement and anticipation in the air, the atmosphere was nonetheless solemn. From a strict-looking woman with long gray hair pulled into an imposing bun, we received a lengthy lecture on the honor and responsibility of being faithful to Lenin's heritage and serving as a shining example for our younger peers who had not yet become *Oktyabryata*.

It would be more accurate to describe us—and by *us* I mean both the 7- or 8-year-olds who, like me, joined the ranks of the Communist children's organization and all those living today, one hundred years after the Revolution— not as the children but as the orphans of October. What do I mean by that?

On the one hand, we are orphaned in time. Thanks to a two-week difference between the old manner of counting days and the new, which resulted from the implementation of the Gregorian calendar in Soviet Russia, the Revolution that happened on October 25 is celebrated on November 7, "a red day in the calendar" as a popular poem dubs it. An October event is not in October; it dislodges itself from the linear succession of time and stands out in its unstable singularity. How to follow it, if it does not coincide with itself or with chronologies defined by calendars and objective time counts? How to catch up with it, or to be born of it, if it runs ahead or lags behind itself?

On the other hand, we are the orphans of October because we deny or are unaware of our filiation. The Russian Revolution of 1917 is generally seen, both inside Russia and outside its boundaries, as a historical aberration, an unfortunate detour on the path of the country's development from a nearly feudal agrarian mode of production to the industrial capitalist regime. Whatever the faults (not to mention the nightmares: the expression *orphans of October* originally applied to the children of people killed in Stalin's bloody purges of the 1930s) that became apparent "the morning after" this event, it voiced the need for real equality, as opposed to the demand of formal equality the French Revolution spearheaded. This need has not found fulfillment; on the contrary, it is more and more frustrated now that the United States and the West as a whole are entering a new Gilded Age of abyssal economic disparities between the richest few and everyone else. Silencing the call for substantive equality that resounded around the world in 1917, precisely when "the world" was engulfed in the First World War, we disinherit ourselves from revolutionary legacy and become its orphans.

In the meanwhile, at the Revolution Museum, I received a small pin with a red star featuring the image of a cherub-faced, curly-haired, blond child (Lenin) at its center. Henceforth, this was a badge to be worn daily on my indigo-colored school uniform. It is easy to criticize the idolatry and the cult of personality that the star of *Oktyabryata* promoted. But from what standpoint are we, the orphans of October, passing such a judgment?

Following the collapse of the Soviet Union, in the 1990s, our global orphanage (rather than a global village) disavowed the revolutionary drive in the name of neoliberalism and its apparent triumph. Since then, on the hundredth' anniversary of the October Revolution, we are faced with a choice between neofascism—replete with parochialism, racism, hysterical nationalism and other such features—and technocracy, as a more ideologically modest heir to neoliberalism. This disastrous situation is the outcome of the historical failure of the revolution and, at the same time, of *our* failure to heed the demand that gave the revolution its impetus. Caught between two terrible options, the one infinitely mirroring and feeding off the other, stuck between neofascism and neo-neoliberalism, we are still living through the fiasco of the Russian Revolution and suffering in the absence of the third (and the only acceptable) alternative of neocommunism.

When the ceremony at the Revolution Museum drew to a close, everyone hurried to join the Revolution Parade. As we passed through the Red Square, Gorbachev and members of the Politburo waived from the grandstand of Lenin's Mausoleum. Standing above the embalmed, museumized body of the leader of the October Revolution, seventy years to the day after it happened, the reform-minded Party Chairman was bent on reinvigorating the political and economic system that, in its institutional form, had embalmed its very *raison d'être*.

We, the orphans of October, should be also fed up with leaning over and mourning the fragmented corpse of the left. What is required now, thirty years later, is a worldwide *perestroika*—of the shattered and defeated left, as much as of the theoretical and practical meaning of the world, of world order, and of being-worldwide. A good place to start is by revisiting the words of the French *L'Internationale*, lost in the standard English translations: "*Foule esclave, debout, debout / Le monde va changer de base / Nous ne sommes rien, soyons tout*" (Stand up, stand up, you enslaved masses / The foundations of the world will change / Let us, who are nothing, be everything). Forget "another world is possible" that, by means of an abstract possibility, prevents actual and necessary change! The world has already changed its foundations, and it is up to the left to realize what has happened and to act accordingly.

4

INCENDIARY WORDS AND THE VOLCANO OF OCCUPATION

At the tail-end of the Third Palestinian Intifada, also known as the Silent Intifada, or Jerusalem Unrest, there were a number of shooting and stabbing incidents on Jerusalem buses, targeting their passengers. This text was written in the aftermath of one such attack, which took place on October 13, 2015.

Visiting the scene of Tuesday's attack on a Jerusalem bus, the city's mayor, Nir Barkat, attributed this and similar incidents to "inflammatory incitement" coming from "mosques and Palestinian leaders." Such superficial explanations are common in today's Israel: just two days ago, Netanyahu himself demanded an official investigation of an Arab-Israeli lawmaker, Hanin Zoabi, for alleged incitement.

The double reference to fire in the mayor's tautological expression ("inflammatory incitement") falls squarely within the field of what I call *pyropolitics*. When a sense of instability supplants the politics of normalcy, I argue, we experience a shift from the politics of the earth, or geopolitics *proper*, to the politics of fire. The new Intifada in Israel/Palestine is, undoubtedly, a pyropolitical irruption on the scene of security for the privileged few, protected by concrete fences and by the ideological façades of Israeli occupation.

Barkat's words are, nonetheless, highly misleading. Operating under the presumption of geopolitics, in which violent, "fiery" outbursts are an exception rather than a rule, he put the stress on discourse and ideology, rather than the unbearable situation *on the ground* (another geopolitical metaphor!) experienced by a vast majority of Palestinian people. From his point of view, the problem is incitement in speech, the use of language capable of fanning the flames of revolt. Such a standpoint denies the victims of occupation their own agency, not to mention the exercise of thinking, or, at least, of assessing whether or not the mode of existence inflicted upon them is tolerable or endurable. "Inflammatory incitement" twice overwrites the culpability of the Israeli authorities and the idea that the material conditions in the West Bank and Gaza have become incompatible with life itself.

Insofar as the protracted occupation of Palestine is instituted in a permanent state of exception, the geopolitical situation in the region is *a priori* pyropolitical. The normalcy of the Israeli enclave has been predicated upon the prevailing and intensifying upheaval and daily disruptions in the lives of the population living and dying under occupation.

As Jewish pacifist-anarchist Nathan Chofshi noted already in 1946, the Zionist movement was building its projected state on a "volcano" of violence, demanding "endless suffering and bloodshed."[1] The long trail of carnage it has unleashed is with us to this day, and it shows no signs of abating. Building on a volcano, in turn, is a very precise image interrelating geo- and pyropolitics in Israel/Palestine. Ostensibly stable political structures are constructed on a shaky foundation of land grabs, disregard for basic human rights, and flouting of international law.

The stabbings that have recently become a daily reality in Israel/Palestine have been widely framed by the international media as an escalation of violence in the region. Before using such charged terms to describe desperate resistance by a people living under occupation, however, it would be prudent to examine these acts in their context. That is to say, it would be advisable to understand them against the background of a seemingly endless, efficiently organized, and institutionalized aggression perpetrated by the Israeli state apparatus and unleashed against the entire Palestinian population.

Indeed, stabbing attacks make violence visible, spectacularize it, but, in doing so, do not produce it as though out of nothing. Rather, they draw the spectators' attention to the already existing, silent, and relentless violence that drives the "attackers" to the breaking point. Their actions are suicidal: they know that, most likely, they will be shot dead in the altercation. Their knifes are powerless in the face of guns that await them at every corner, now that Jerusalem's mayor has called on the Jewish residents of the city to carry registered firearms with them at all times.

Such a technological imbalance is typical of colonial occupation, as already noted by Frantz Fanon in the 1950s and 1960s. And the same asymmetry affects the numbers of those who succumb to violence on both sides; disproportionately more Palestinians than Israelis die in the altercations. The victims of stabbings are individuated, each with her or his story; the victims of occupation are not only more numerous but also anonymous, unnoticed, submitted to the additional violence of remaining nameless.

It is to be expected that reactive violence would flare up once attempts at a dialogue have failed. Hannah Arendt expressed this point well in her political-philosophical body of work: when speech, or *logos*, is brushed aside, mute physical aggression takes its place. The absolute unwillingness of the current Israeli government to engage in a meaningful dialogue with its Palestinian neighbors, with the view to urgently ending occupation and creating a viable Palestinian state, is at the root of the language of force that, alone, is speaking today.

The active violence, to which the current incidents are a response, is the grinding violence of occupation: severe travel restrictions, land grabs through the construction of the West Bank barrier considered illegal by the UN, arbitrary arrests, the destruction of houses, be it by bombings or demolition orders... Let us pause and think, instead of automatically reproducing the ideologically inflected jargon of the mass media. Is the "cycle of violence" limited to stabbings and revenge attacks on both sides? Or is it encrusted into a larger and particularly vicious circle—the suffocating siege of a people denied the basic conditions for living and a chance for their collective self-expression?

Violence is not an effect of fiery speeches—the "inflammatory incitement" blamed by Barkat—but an eruption of the volcano that is Israeli politics-as-usual. For once, it is crucial to understand that the construction of the world in words has its limit, which is the unutterable, unspeakable, and intolerable theft of everything that makes life materially possible (land, water, movement, nourishment, medical treatment, livelihoods). And that no fiery words can compete with the volcanic politics of occupation.

5

CAN THERE BE POETRY AFTER NETANYAHU?

In early 2016, the right-wing group *Im Tirtzu* led a vicious public campaign against the Israeli artistic community, whose members, in their majority, fall on the leftist end of the political spectrum. The artists were presented as traitors to their country. Despite the official condemnation of this campaign, the Israeli Culture Ministry sought to allocate state funding only to those groups of artists who were "loyal to the state."

Distance themselves from the new *Im Tirtzu* (If You Wish) smear campaign waged against left-wing artists in Israel as they might, Netanyahu and his government are thoroughly complicit with the efforts to brand their political opponents and artists opposed to the status quo as traitors.

The tactic is nothing new. It was developed and honed in the months leading up to the assassination of Prime Minister Yitzhak Rabin in mass right-wing rallies—where Rabin was depicted with Hitler's moustache and uniform—attended and addressed by Netanyahu. Then, there is the more recent government bill, demanding that the Israeli NGOs, recipients of funding from abroad, make a public declaration to that effect and face strict regulations as a result. Finally, and more recently, the Israeli culture and sports minister Miri Regev tabled a proposal for a "Loyalty Bill" that would allocate government grants solely to artists supportive of the regime. So, aren't there good reasons to be skeptical regarding Netanyahu's comments that he does not view his political opponents as traitors?

In the mainstream media, the proposed Israeli legislation has been likened to the McCarthyist witch hunts in the United States of the 1950s. The terms of comparison are quite inevitable in light of the admiration the founder of *Im Tirtzu* has publicly expressed for Senator McCarthy. But we need not travel so far in time to find alarming parallels. It is obvious that the Israeli right is emulating above all Vladimir Putin's Russia of the 2010s. Putin, too, considers the liberal opposition to his "vertical of power" treacherous, and its representatives—the agents of the West, bent on destabilizing the country by undermining its social cohesion, of which he, presumably, is the only guarantor. He, too, has endorsed a law that potentially prevents any organization relying on funding from abroad from operating on the territory of the Russian Federation. And he, too, has clamped down on renegade artists, especially in the wake of the international success of Andrey Zvyagintzev's feature film *The Leviathan* (2014), partially funded from state coffers and highly critical of rampant political and economic corruption.

The analogies between the measures being introduced in the two countries are not accidental. Former residents of post-Soviet states comprise over 20 percent of Israel's population. Likewise, the political objectives of Putin and Netanyahu dovetail: colonial expansion of the respective territories they govern is high on the agenda. Both leaders are obsessed with "internal enemies," the critics whom they deem to be collaborating with hostile outside forces. They both strive to hold onto a nineteenth-century version of nationalism, outdated virtually everywhere else in the West and based on the construction of a homogeneous political entity.

Among all these similarities, however, we should not overlook some glaring differences to do with the divergent historical contexts of Putin's and Netanyahu's rule. The former is an inheritor of Soviet censorship, where the political repression of those artists who did not fall in line with the regime was rife. I am referring not only to the heyday of Stalinism but also to the so-called thaw of Khrushchev's period, when Joseph Brodsky stood trial for "parasitism" and was condemned to hard labor in Siberia. Along with, before, and after Brodsky, who will later go on to win the Nobel Prize for Literature, many other Jewish writers and artists suffered at the hands of political authorities they did not please. In Nazi Germany, for instance, Jewish art and thought were labeled *degenerate* as a stand-in for the innovative, avant-garde trends of modernism.

In the face of the painful—indeed, tragic—historical experience of Jewish artists, their freedom within the State of Israel should have been sacrosanct. Similarly inviolable in the eyes of the state should have been the artistic freedom and freedom of thought of non-Jewish residents, given the clear resemblances between their position and that of countless diasporic Jews, practitioners of "the creative occupations." Philosopher and cultural critic Theodor Adorno once said that "to write poetry after Auschwitz is barbaric."[1] Perhaps. But I also have no doubt that more barbaric still is to persecute certain writers of poetry, novels, drama, music, and so forth in the name of a state that has emerged "after Auschwitz."

6

MARGINALIZING EUROPE

Written in the beginning of 2013, this essay considers the trend of "colonization in reverse," in which the economic elites from the non-Western world purchase majority shares in the newly privatized or privatizing companies and financial institutions in the West.

An unintended consequence of the current economic and political crises in Europe has been the completion of the Continent's decolonization, commenced in the middle of the twentieth century. As the gross domestic products of developing countries continue to grow, while many crisis-stricken EU economies are contracting, some of the formerly colonized nations, alongside China, are actively purchasing the assets that are being privatized in Europe.

The Angolan media group Newshold is preparing to purchase the public television channel RTP in its old colonial master, Portugal. EDP (*Energias de Portugal*), another Portuguese company that generates, supplies, and distributes electricity on the Iberian Peninsula ceased being publicly owned, when 21 percent of the state's stake in it was sold to the China Three-Gorges Corp. OPAP, a highly profitable Greek gambling company, similarly received privatization bids from a Chinese corporation. Pireus Container Terminal in the famous Greek port is now a subsidiary of the Chinese company Costco. Nor should we forget the inverse trend of de-privatization outside Europe: in 2012, Argentina nationalized the Spanish-controlled oil producer YPF at the same time as Bolivia seized the subsidiary of the Spanish group *Red Eléctrica Corporación* based in the South American country.

At issue is not merely the privatization of profitable public enterprises all over Southern Europe, complemented by the nationalization of companies nearing collapse, such as *Bankia* in Spain. The above examples amply demonstrate that it is no longer necessary to "provincialize Europe," as the pioneers of postcolonial studies had recommended, because Europe is doing an excellent job provincializing itself!

As the internal squabbles among EU member states continue and as the crisis becomes more protracted in the absence of systemic reforms and regulation overhauls, Europe risks losing its position of a power hub in a multipolar world. The zero-sum mentality of the still prospering European states blinds them to the fact that, should the countries on the periphery be demoted to the status of second-class members, the entire Union would be dragged down along with them. From a dominant colonial center, Europe is quickly turning not into an equal partner of Asia and of the states it had colonized in the past, but into their subservient outgrowth.

The movements of capital are also mirrored by the trajectories of human migration. Emigration from Portugal to Brazil and Angola skyrocketed in the last few years: between 2009 and 2010 alone, there was an increase of 60,000 in the number of Portuguese citizens registered at consulates in Brazil. The number of people, who emigrated from Spain in the first six months of 2012, saw a 44 percent increase, compared to the same period in the previous year. Although 86 percent of these were naturalized Spanish immigrants born overseas, who decided to return to their home countries in Latin America where economic conditions are improving, a majority of Spaniards have expressed their willingness to live abroad, if work opportunities there presented themselves.

From the postcolonial point of view, Europe requires a healthy dose of marginalization, which would act as a remedy against the centuries-old Eurocentric prejudice and claims to cultural superiority. But, before celebrating this "turning of the tables" on the former colonizers, it would be prudent to ask if a more fundamental shift has really taken place.

As the postcolonial era was ushered in, the promise of a more just economic arrangement was sadly unrealized. National capitals simply changed hands from the colonizers to small local elites, often with close family and other ties to the metropole. The vast majority of populations in the postcolony were denied better standards of living, while the natural resources of the newly created countries continued to be plundered and sold so as to fill private purses.

The EU crisis is presenting a distinct opportunity for key players in the emerging economies to purchase public (and quite successful) European companies at relatively low prices. Coupled with the exacerbation of the internal inequalities between the center and the periphery of the EU itself, the completion of decolonization threatens to pass, seamlessly, into a reverse colonization by purely economic means.

A weakened, marginalized Europe does not correspond either to the best interests of its citizens or to the stability of a multipolar global political order. Although a paradigm shift toward the decentering of European collective consciousness is to be welcomed, this change should be accompanied by a continued insistence on the universally binding ideals of social and economic justice historically championed by Continental thinkers and political movements.

7

THE EUROPEAN UNION AND THE RHETORIC OF IMMATURITY

On February 16, 2012, the then head of the Eurogroup, Jean-Claude Juncker, chastised Greece for its lax implementation of the austerity measures demanded by the Troika and insinuated that the peripheral and southern member states required more external oversight and supervision.

Over two hundred years after the death of Immanuel Kant, the question of political enlightenment is still a hot issue in Europe. What does it take to attain both individual and collective autonomy and the capacity for self-governance? Who has managed to throw away the shackles of "self-incurred immaturity," as Kant famously put it, and to be in full control of oneself, as well as of one's public affairs?

The European Union operates on the assumption that certain member states with excessive amounts of unsustainable sovereign debt are not as enlightened as others. Greece, for instance, has been submitted to a series of humiliations, including proposals for the EU to take direct control of its state budget. On February 16, 2012, the head of the Eurogroup Jean-Claude Juncker echoed this sentiment, suggesting that the implementation of austerity measures in Greece is in need of "surveillance" and "tighter oversight." In other words, the leadership of the EU thinks that Greek people are collectively childish enough to require adult supervision in order to ensure that they spend the bailout package in the most responsible way possible.

The tendency to infantilize select member states is in line with their animalization, evident in the insulting abbreviation of Portugal, Ireland, Greece, and Spain in the word PIGS, neither as human nor as rational as the rest of the EU countries. Throughout Western philosophy, both children and animals, with their capricious wills, have been considered deficient from the standpoint of fully developed rational adults and, hence, in need of training, education, and disciplining. The current austerity measures are, in fact, a version of collective punishment, inflicted not so much to control spending and improve economic performance (all indicators clearly show that economic conditions are worsening in countries where such measures have been implemented), but to force the vast majority of citizenry into submission, poverty, and willingness to work for absurdly low wages. Following the logic of Christianity, to which Kant also undeniably subscribed, this veritable collective self-sacrifice would serve as the motor of progress, if not of full humanization, when it comes to childish or animal-like nations.

The historical process we are going through is not comparable to the Great Depression but, rather, to the transition from the Speenhamland System (1795) to the Poor Law Amendment Act of 1834 in England, when earlier welfare provisions were repealed in favor of legislation that endeavored to convert the British poor into "hungry animals." The new laws purposefully utilized hunger as a disciplinary mechanism to force people to work for meager wages. As Joseph Townsend put it in his *Dissertation*, the ideological cornerstone of the 1834 Act:

> Hunger will tame the fiercest animals, it will teach decency and civility, obedience and subjection, to the most perverse. In general it is only hunger which can spur and goad them [the poor] on to labour; yet our laws have said they shall never hunger [...] [L]egal constraint is attended with much trouble, violence, and noise [...] whereas hunger is not only peaceable, silent, unremitting pressure, but as the most powerful natural motive to industry and labour, it calls for the most powerful exertions.[1]

Fast forward to 2012. The shameful efforts to reduce wages in the name of economic competitiveness through increased taxation and cuts in Greece, Portugal, and a number of other EU countries sound like a distant echo of 1834. Then, like now, whole social groups and political-economic classes were dehumanized, treated as animals or children to be pacified through the physiological mechanism of starvation. The exacted fiscal discipline, which entails, among other things, shocking cuts to pensions and unemployment benefits, is the kind of disciplining that lacerates the entire *body politic* in the name of calculative rationality and maximization of efficiency—the key ideals of "enlightened" thought. (We would not be surprised to learn that, in the aftermath of passing new Poor Laws in the nineteenth century, England erupted into riots, just as Greece did in the twenty-first century, after the approval of the new bailout agreement by its Parliament.)

The infantilization and animalization of entire nations is, of course, nothing new for Europe, which has had a long tradition of portraying itself as the beacon of humanity and which has invariably resorted to the idea of its "civilizing mission" throughout colonial conquests and expansions. Now, almost four decades after the last European countries have withdrawn from the colonies overseas, the same rhetoric is being turned inward, retracing the new political-economic rift between the North and the South of Europe. Exploitation is the one constant that remains after this shift: exorbitant interest rates and repayment conditions attached to bailout packages will guarantee that the debtor countries organize their economies around the need to service their debt for the foreseeable future.

The bitter irony is that demands for fiscal restraints and rational self-control in the economic realm are posed before governments but not before multinational corporations, banks, or hedge funds. Indeed, transparent oversight of these economic actors has been gradually eroded and reduced, resulting in the global economic crisis, for which ordinary citizens are now expected to pay with their very livelihoods. Irresponsible gambles with derivatives and other nonexistent goods on the global markets do not fall under the heading of unenlightened immaturity, while overspending meant to improve social services, healthcare, and social safety nets is seen as irrational. The self-professed guardians of the European legacy fall far short of the Kantian ideal they tout. It is now time to turn the rhetoric of immaturity against them, so as to reveal the double standards and the capriciousness of their logic.

8

TRUMP METAPHYSICS

In the Republican primaries of 2016, Donald Trump came under attack by others seeking the Republican nomination to the presidential ticket. In a grotesque way, the matters debated on the sidelines of the primaries resonated with the key questions of Western philosophy.

Quite unexpectedly, a bunch of philosophical—I would even say "metaphysical"—issues have come up on the Republican side of the primaries this election season. How to distinguish the real from the fake? What level of ignorance is simply unacceptable in public affairs? How to espy matters of principle, or something like "the inner essence," behind changing appearances? These questions have been, in one way or another, the staples of Western philosophy ever since its inception. And most of them have been now linked to the candidacy of Donald Trump.

Take, for instance, the recent GOP debate organized and aired by Fox News. Presenter Megyn Kelly quizzed Trump: "The point I'm going for is: you change your tune on so many things, and that has some people saying, what is his core?" Her point goes beyond the usual flip-flopping accusations leveled against presidential candidates. It even overflows the opposition between a politics based on immutable principles, often called *the politics of truth*, and an opportunistic catering to various groups comprising the electorate. Unwittingly, having thrown anything but the kitchen sink at Trump, Kelly has dug up a crucial metaphysical distinction between the stable, selfsame inner essence and fleetingly superficial outward appearances. In his response, Trump insisted that there was no contradiction between his "very strong core," upon which he did not elaborate further, and "a certain degree of flexibility" necessary for learning from past experiences. His inaccessible essence thus reconciled with evanescent appearances, Trump gave himself a meta-excuse for any and all crude inconsistencies in his take on domestic and foreign policies alike.

Along similar lines, Mitt Romney's March 3 verbal assault on the current Republican frontrunner touches on an issue dear to the philosophical heart. The failed 2012 candidate called Trump "a phony, a fraud," as well as "a con man, a fake." Since its inception in ancient Greece, philosophy, too, has been suspicious of an oratory that substituted a flowery or a fiery rhetoric for the things themselves. Plato's *Republic* associated the political sphere as a whole with such empty and deceptively manipulative strategies, while prescribing a universally valid method for leaving the cave of appearances with the assistance of the philosopher-king. But before identifying Romney with a modern-day (latter-day) Plato or Socrates, we ought to inquire: in the name of what truth is he condemning Trump? The critic overtly assumes that there are Republican politicians who are not fake, those authentically suffused with the bracing tenets of the "conservative movement." Brushing aside Romney's assault, Trump characteristically turned the tables on him and reminded voters of how Romney begged for his support as he was running against Barack Obama. Obviously, the accusation "you are a phony, a fraud" loses much of its bite if it comes from someone revealed to be a phony and a fraud in a field populated by similar phonies and frauds.

It is simply futile to chastise Trump from the standpoint of stale metaphysical values, because he embodies a system, which has a long time ago outgrown and abandoned these same values. What does it mean to decry a candidate to the office of president as a "fake" in a country where a Hollywood actor was president (more precisely, enacted the role of president) for two consecutive terms? Does it make sense to bemoan this candidate's ignorance less than eight years after the end of George W. Bush's terms in office? Where is the logic of accusing him of vulgarity when the official pick of the Republican establishment for the presidential race hints at differences in penis sizes as momentous for the outcome of the contest?

The reason behind the fact that Trump is currently leading (in a dismal field, mind you) is not, as Linda Martín Alcoff has argued in *The Philosophical Salon*,[1] that his own ignorance appeals to certain ignorant white voters. Or, at least, it is not *the only* reason. Rather, what Trump does most deftly, and what in my view accounts for much of his current success, is that he fully assumes the bankruptcy of the metaphysical ideals such as authenticity, essentiality, or firm principles, and acts accordingly. His rivals, in turn, are aware of the crumbling of metaphysics but pretend that it is still alive and well. In both cases, nothing supplants the outdated value system, except for self-serving private interests and megalomaniac aspirations.

Curtly put, the bygone values are supplanted by nothing—by the nothing, to which everything has been reduced. Whereas Ted Cruz & Co. stand for the consciousness of this nothingness, Trump represents its self-consciousness, and this gives him an unmistakable edge over his rivals. He knows how to use the pure nothing that he represents, even as the other presidential contenders pretend that there is something behind *their* nothing. And so, Trump comes across as much more authentic in his inauthenticity than the others, who are busy drawing, in Plato's words, the "shadow paintings of virtue" all around themselves.

A deeper cause for the GOP establishment's concern and dissatisfaction with Trump is that he puts a mirror before it, forcing it to face up to its disavowed reality and exacerbating its tendencies in the process. In order to dissimulate the unpleasant truth, the party has no other choice but to distance itself from the rogue candidate, who uses even this lack of official support for his bid to his advantage, as proof of his outsider status, his non-belonging in the world of "Beltway politics." Any attack can be turned around to serve Trump's purposes, especially if he is censured based on the precepts of metaphysics, which have long become those of "common sense." *Trump trumps metaphysics*: herein lies the recipe to his success so far in the campaign. To oppose him better and more effectively, we would need not to recycle stale metaphysical slogans but to chart other paths towards what lies beyond metaphysics. Towards a multiplicity free of totalization, a proliferation of differences, and a sense of sharing that has dispensed with the very idea of property.

9

THE CON ARTISTRY OF THE DEAL: TRUMP, THE REALITY TV PRESIDENT

This essay is a prospective look at the Trump presidency, cast right after the 2016 US elections.

Four days prior to the US presidential elections, I read Trump's *The Art of the Deal*—a manual for con artists and, as it turned out, an updated version of Machiavelli's *The Prince* for the media age. Having dismissed the Trump candidacy early on in the primaries season as a marketing gimmick intended to promote his overall brand, I wanted to avoid making the same mistake at the very final stages of the campaign. The same cannot, unfortunately, be said of Clinton's team, which clearly did not do its homework. Were her advisors at least to skim through *The Art of the Deal*, they would have promptly realized that no amount of negative publicity could damage the Republican candidate, for whom there is no such thing as "bad" advertisement. Whatever the context, to be mentioned 24/7 on cable news is, for Trump, a goal in itself, making him larger than life and, therefore, an ideal in which common folk can espy their own unattainable dreams and desires of grandeur.

The most emblematic passage from the con artist's manual, and one most relevant to Trump's political strategy, is the following:

The final key to the way I promote is bravado. I play to people's fantasies. People may not always think big themselves, but they can still get very excited by those who do. That's why a little hyperbole never hurts. People want to believe that something is the biggest and the greatest and the most spectacular. I call it truthful hyperbole. It's an innocent form of exaggeration—and a very effective form of promotion.[1]

37

Isn't this precisely what Trump did throughout the presidential campaign? Did he not play to people's fantasies, be they about the revival of the country's industrial might or, on the darker side, about the possibility of achieving social purity by excluding every kind of threatening Other? Far from "innocent," the exaggerations in question helped Trump close the deal with the American public at the price of unleashing a whole range of sexist, racist, nationalist, and homophobic fantasies. It bears mentioning that these fantasies were not created by Trump, but merely driven out of hiding in the deep recesses of the unconscious, where they had been taking refuge from the superficial dictates of political correctness. Positively buttressed by projective identification with the "greatest and the most spectacular" business mogul, the phantasmic construction of "America's greatness" put the self-professed artist of deals in the Oval Office.

The United States already had its actor president in Ronald Reagan. And it is about to get its reality TV president in Donald Trump. What this means is not only that the last dividing lines between fantasy and reality are being erased but also that fantasy itself, however ominous and disturbing, comes to determine reality.

To paraphrase Plato, Trump is going to be the sophist-king, a master puppeteer of appearances who knows how to manipulate them and to manipulate their very manipulation (e.g., by lambasting the mass media that have literally made him what he is today). As I mentioned in "Trump Metaphysics," the manipulation of public opinion is nothing new; rather, what distinguishes the president-elect from his rivals, including Hillary Clinton, is that "he fully assumes the bankruptcy of metaphysical ideals such as authenticity, essentiality, or firm principles, and acts accordingly..."

Instead of giving in to despair in the face of the current political success of con artistry (after all, a strong tradition exists in philosophy, according to which politics *is* con artistry; Plato uses this conclusion to denounce the political realm of appearances as a whole, while Machiavelli builds upon it to formulate the guidelines for successful politicians), I'd like to highlight two of its inherent limitations and one unexpected positive implication.

First limitation. Note that, in the case of Trump's political ascent, the medium is not the message. The form he resorts to is decidedly postmodern: the mediatic construction of reality, recently theorized in William Egginton and David Castillo's fascinating book *Medialogies*.[2] But the content is essentially modern, embracing exclusionary nationalism and industrialism. Inevitably, that contradiction, the power of which Trump has harnessed and which he has been riding thus far, will be resolved in favor of one or the other extreme, contributing to the process of post-electoral normalization. What will happen when @RealDonaldTrump (we can't overlook the irony of "Real" in the Twitter handle) switches to @POTUS? When the political movement that brought Trump to power culminates in the static and homogeneous—all-male, all-white—Politburo-like structure of governance? When the new administration becomes one of the least transparent, most secretive in US history?

Second limitation. All the self-contradictory promises the Trump campaign has made will have to give way to actual policy decisions, starting with the choice of cabinet, which doesn't at all look like it will be comprised of political outsiders. As Hegel reminds us in *Phenomenology of Spirit*, possibilities, capacities, talents, inclinations, and so forth, are abstract and, because they are not yet realized, admit everything into their ambit. With their realization, however, certain possibilities fall by the wayside, insofar as they are not acted upon, or insofar as they stand diametrically opposed to those actualized. As a result, another crucial "Trump card," namely his resistance to being pinned down to one concrete position, will be lost. The author of *The Art of the Deal* is aware of this mode of taming the otherwise unrestrained fantasy: "You can't con people, at least not for long. You can create excitement, you can do wonderful promotion and get all kinds of press, and you can throw in a little hyperbole. But if you don't deliver the goods, people will eventually catch on."³ The point is that, given the self-contradictions he has been cultivating, Trump will have no other choice but to fail in "delivering the goods," even if he tries to do so. The very strategy that got him elected will backfire in the period of his presidency; it is one thing to break definite election promises, but it is quite another to break promises in fulfilling them.

The *positive implication* of the Trump presidency resonates with Slavoj Žižek's recent analysis of its prospects.⁴ Commentators are up in arms that Trump's stint at the helm of the United States will spell out "disaster for the planet" and an assured defeat in the battle against climate change,⁵ disaster for the most vulnerable and the poor, disaster for race relations … They forget that the Paris climate agreement is too little, too late to keep the world livable, or that wages have not increased and that race relations have hit a new low under the Obama presidency. What the election of Trump signals is that the ideological screens concealing these and other unmitigated catastrophes have fallen and that we can no longer congratulate ourselves on symbolic victories, while moving at full speed toward environmental and social breakdowns. This is how fiction realized makes disavowed reality itself real.

10

COVID-19: THIS IS NOT A WAR

When the COVID-19 pandemic first erupted, politicians around the world adopted a militaristic discourse with regard to the virus. I made the appeal to refrain from this way of speaking about the virus in March 2020. The tendency is continuing into 2021.

About the current coronavirus pandemic and a concerted response to it, we should say unequivocally: "This is not a war." True, such a statement would directly contradict the words of many heads of state who have declared a war on the virus. To deny that a militaristic framing is necessary, or even adequate, is not to close our eyes to how critical the situation is. It is, instead, to search for alternative ways of grappling with the situation, of inspiring people for collective and individual action, and, ultimately, of bringing about a better world after the pandemic winds down.

1. *Prehistory of "the war on virus."* Since the 1960s, governments around the world (beginning with the United States) have been extending the discourse of war beyond the context of military hostilities, as traditionally understood. In 1964, US President Lyndon Johnson announced the start of a war on poverty as he attempted to institute a welfare state in the country. In 1971, President Richard Nixon called drug abuse "public enemy number one" and declared a war on drugs. In 2001, President George W. Bush, sounded his call for a global war on terror in response to the 9/11 attacks on the World Trade Center in New York. The 2020 war on the coronavirus should be seen in the context of these declarations.
2. *Invisible enemy.* With each new declaration, the presumed enemy became more and more invisible, lacking recognizable outlines. It—rather than she or he—could be just about anywhere. With the enemy not easily

localizable and potentially ever-present, war became total, engulfing reality as a whole.

3. *The logic of war.* The invisible enemy that figures in a war on the coronavirus totalizes war by erasing a clear front line. While the line is erased, the front does not disappear: it is drawn between and even within each of us, given the uncertainty of whether or not one is infected with the coronavirus. Another element of war that becomes distorted in present circumstances is the real possibility of killing and being killed. Neither the virus itself nor those it infects have the *intention* of killing anyone. So, in a war paradigm, the role of the virus is ambiguous: is it an enemy or a weapon? Is a potentially infected human body the virus's weapon? Or is it, itself, an enemy? Leaders who fall back on militaristic metaphors have the responsibility of thinking through the logic they operate with and its consequences.

4. *Victory.* In wars extended beyond the sphere of armed conflicts between human groupings, victory is unattainable. Neither is defeat. Not only do wars on drugs, terror, and now a virus become all-embracing; not only do they erase the front line and a discernible enemy figure, but they also have no end in sight, no definite cessation of hostilities. An inflated concept of war runs the risk of becoming a fight for a cause lost from the get-go.

5. *Peace.* Assuming that one could declare one's victory or admit to being defeated in such wars, what would the peacetime that follows look like? In fact, peace is not at all contemplated in hostilities against terror or a virus. The maximalist objective they have is the complete elimination of the enemy, its total annihilation. These are wars without peace and, therefore, without the opposite that would limit them, in time or in conceptual space.

6. *Destruction of the common good.* After decades of neoliberal policies that have resulted in the privatization of utility companies and pension funds, the erosion of workers' rights, divestment from public healthcare and other vital sectors and services, the experience and the notion of the common good have been rendered hollow. As a result, appeals to a population to act for the common good will fall on deaf ears and will not produce the same desired, emotionally charged effects as a declaration of war, implying the need to mobilize, to combine individual efforts and to make sacrifices.

7. *A unique opportunity*. Terrifying and tragic as it is, the coronavirus pandemic presents a unique opportunity—to rebuild the sense of the common good. We need to concentrate on the small acts of kindness and solidarity all around us: people offering older neighbors their help with buying food provisions or medicines; caring about others (so that they would not go without and would not become infected); not to mention the enormous risks that medical personnel take in treating those who have contracted the virus. Combined with some government actions, such as the abolition of the difference between public and private healthcare systems, these experiences may reinvigorate the notion of the common good. If an appeal to this notion were to make sense again, then it would be significantly more effective for overcoming an emergency situation than the frames of war that are again being thrust upon us.

11

GOING VIRAL, OR
THE CORONAVIRUS IS US

Right before COVID-19 was declared a pandemic by the World Health Organization, I penned the following analysis for *The New York Times*.

The new coronavirus outbreak, now threatening to snowball into a full-fledged pandemic, should provide us with an occasion for thinking. What is the nature of power and sovereignty in an increasingly interconnected world? How do borders, membranes, boundaries, and limits—of individual bodies or of entire communities—function? In what senses does physiology mold and is molded by social and technological realities? And how do viruses challenge the traditional views on governmentality, individuality, and life?

Well before the current outbreak, the global propensity has been to build walls and seal off national borders—between the United States and Mexico, Israel and Palestine, Hungary and Serbia and Croatia, among other places. The resurgent nationalism that had instigated this propensity has nourished itself on the fear of migrants and social contagion, while cherishing the impossible ideal of purity within the walled polity.

Border closures and quarantines in response to the new coronavirus are largely symbolic measures stemming from the same basic logic as the construction of political walls: both acts are meant to reassure citizens and give them a false sense of security. At the same time, the proponents of such measures turn a blind eye to the main problem, namely the underdeveloped mechanisms of transnational cooperation, governance, and decision-making that are vitally important for tackling climate and migrant crises, pandemics, and tax evasion.

Survivalism has always followed a trajectory parallel to that of virulent nationalism. At its core is the fiction of a self-reliant, totally independent and autonomous, Crusoesque individual, the one who is smart and strong enough to be able to save himself (the gendered pronoun is not accidental here) and, perhaps, his family. Following in the trail of the theological doctrine of salvation reserved for the select few, this attitude abstracts human beings from the environmental, communal, economic, and other contexts of their lives. As panic sets in in some quarters, personal border closure imitates the knee-jerk political gesture: food and medical supplies are hoarded, while the wealthiest few prepare their luxury doomsday bunkers. What, in contrast to such fictions, the novel coronavirus demonstrates is that borders are porous by definition; no matter how fortified, they are more like living membranes than inorganic walls. An individual or a state that effectively manages to cut itself off from the outside will be as good as dead.

More than occasional threatening eruptions on the otherwise calm global horizons, viruses are the figurations of the contemporary social and political world. (We might say, tongue in cheek: the bug is not a bug but a feature.) A symbol subtler than the wall thus emerges with SARS-CoV-2, the virus that triggers COVID-19—the symbol of a crown. SARS-CoV-2 belongs to a group of RNA viruses transmissible between animals and people. As this feature indicates, it does not obey natural systems of classification and species boundaries, either. The club-shaped spikes on the virus's spherical surface have earned it the name *coronavirus*, derived from the Latin word for "crown," *corona*, which harkens back to the Greek *korōnē*, meaning "wreath" or "garland." One of the most recognizable attributes of sovereignty, a crown is bestowed on a microscopic entity that defies distinctions between various classes of beings, let alone between the domains of life and death. Transgressing old borders, the virus becomes a figure of sovereignty in the age of the dispersion of power. If so, then to understand its workings is to get a glimpse of the way power operates today.

One aspect of viral activity is to infiltrate and to transcribe the texts of host cells and computer programs. Another is to replicate itself as widely as possible. In the social media universe, both aspects are actually coveted: when a photograph, a video, a joke, or a story is shared, quickly spreading among Internet or cellphone users, it is said to go viral. A high rate of viral content's replication is not sufficient, as it needs to make an impact, transcribing and hijacking, as it were, the social text it has infiltrated. The ultimate goal is to assert one's influence through a massively disseminated image or story and wield power thanks to it. Going viral introduces a fair degree of complexity into our affective relation to viruses: feared, when we become their targets and possible hosts; desired, when they are our instruments for reaching a sizeable audience.

The comparison of going viral on the Internet and a coronavirus pandemic is not far-fetched. The global dimension of recent epidemics is due to the increasing mobility and physical interconnectedness of large segments of world population engaged in mass tourism, educational and professional exchanges, long-distance relations, international cultural and sports events, and so on. It was on board of cruise ships, such as "Diamond Princess," on planes, in trains and hotels that the virus travelled beyond the hotspot of its initial outbreak—in other words, in those instances when one sent *oneself*, not just one's image or message, elsewhere.

Whether we like it or not, we are all hosts for elements that are alien to us at every level of existence. Moreover, there is always a risk that hosts would be harmed by those they host. This risk is ineliminable. Rather than conjure up the specters of sovereign nation-states and autonomous individuals, we need to learn to live in a world that is interconnected not only ethereally or ideally, through communication technologies, but also materially, via direct embodied contact. In short, we must learn to live in and with a reality that may, at any moment, go viral.

12

CAN DEMOCRACY SAVE THE PLANET?

In the summer of 2017, the Trump administration announced that the United States was withdrawing from the Paris climate change treaty, which it joined under Barack Obama in 2015. In 2021, Joe Biden made a renewed commitment to the Paris Agreement, one of his first executive acts as President of the United States.

Donald Trump's announcement of US withdrawal from the Paris climate change treaty was not unexpected. Both before and after being elected president, he stated his opposition to climate action publicly and unswervingly, online and in campaign speeches. Take, for instance, his notorious tweet from November 6, 2012: "The concept of global warming was created by and for the Chinese in order to make U.S. manufacturing non-competitive."

Why has Trump's position on the issue, unlike his stance on sundry topics, remained so stable over time? This exceptional consistency has nothing to do with principles. What matters is that the evidence for climate change is not (at least, *not yet*) entirely direct and not quite mediatic; it does not provoke an instant sensory shock, as images of children who fell victim to a chemical attack did, presumably prompting Trump to order the bombing of military installations in Syria. Even if the extractive-destructive fossil fuel industry, to which the US President and many key players in his administration are directly linked, were magically to disappear overnight, little would change in his approach to global warming, which requires an intellectual as opposed to a purely sensory-emotive response, the work of imagining events over long time spans as opposed to a knee-jerk reaction to a present stimulus.

In a recent opinion piece, titled "America's Broken Democracy," Jeffrey Sachs attributes the fiasco of the Trump-led disengagement from the Paris agreement to a "deep corruption of the American political system," which "has become a game of powerful corporate interests: tax cuts for the rich, deregulation for mega-polluters, and war and global warming for the rest of the world."[1] It is, nonetheless, unclear when the change from a pure to a corrupt version of politics has taken place on the ground. When has the American political system been at the behest of interests other than those of powerful corporations? Paradoxically, Sachs underwrites Trump's own discourse of the lost greatness of America, though he sees in the current US President the culmination of that loss, not the hope for salvation.

What caught my attention in Sachs's analysis was not his misguided bemoaning of a "once-functioning democracy" in "America," but his implicit assumption that representative democracies at the level of nation-states are the political formations best suited to tackling the planet-wide, anthropogenic environmental disaster we are in the midst of.

In addition to the class-based ideological manipulation of public opinion and electoral choices, actually existing democracies (and we are badly in need of this concept that would be analogous to "actually existing socialism" coined in the twentieth century) operate with ultra-short timeframes and have an extremely narrow spatial reach. Four- or five-year electoral cycles fail to measure up against the centuries of climate change, requiring policies with an equally long-term commitment. The territorial limits of national sovereignty do not give a sense that meaningful environmental action in one country will have significant impact on a global scale. The fickleness of "the people's choice," as far as democratic political leadership is concerned, is more and more reminiscent of televised contests, such as the musical *Idol* competitions now popular everywhere around the world.

It is not that democracy is broken here or there, with political and economic corruption no longer concealed by the screens of ideology. Rather, democracy *is* brokenness. Broken in time, it is out of sync with the *longue durée* of the most pressing, life-and-death environmental problems, particularly air and water pollution, the melting of Arctic ice, rising sea levels, and so forth. Broken in space, it is incapable of addressing these same problems that affect both human and nonhuman beings everywhere on the planet. The Band-Aid solution of international accords, such as the 2015 Paris Agreement, is ineffective, if they remain non-binding and at the mercy of shifting political moods at home.

Yet, if the brokenness of democracy is something we might for better or for worse live with (or even want to live with) for the time to come, then it is paramount to ensure that this political regime does not violate the material conditions for life on earth. While we are at it, we might as well do so in the open, neither by measuring actually existing democracies against an impossible democratic ideal nor by bemoaning their lost Golden Age, which is nothing but a fiction. How then? By reinventing democratic actuality and asserting: *only a planetary democracy can save the planet.*

A planetary democracy would be (1) planet-wide; (2) of the planet; (3) for the planet. These are its minimal parameters that refrain from specifying its form. Indeed, any imposition of a political form in this context would be both ineffective and dictatorial; another actuality of democracy must develop through a democratic discussion, itself responding to the three parameters I have listed. Such a conversation, in turn, must be authorized, promoted, and supported (including financially) by the vanguard of actually existing democracies mindful of their structural incapacities to make any other substantial contribution to a liveable planet.

As I have indicated, I do not think that it is methodologically or ethically right for a philosopher to offer blueprints for a democracy beyond the nation-state and beyond the human, as Michel Serres and Bruno Latour among others, have done. That said, I do consider planetary democracy to be a response to Marx and Engels's exhortation, "Proletarians of all countries, unite!" (*Proletarier aller Länder, vereinigt euch!*) The Latin-derived meaning of the word "proletarian" exceeds the bounds of the human: it refers to the offspring or progeny (*proles*), which was the only service non-propertied ancient Romans could give to their political community, maintaining and increasing its labor power. In this sense, not only humans but also plants and animals, fungi and bacteria, are members of the proletariat projecting themselves into the future through their descendants. Such a tendency accounts, also, for the variation *vegetariat*, which Cate Sandilands has introduced to express how contemporary capitalist exploitation taps into the vegetal capacities for nourishment and reproduction shared by plants, animals, and humans.[2] (Lest an objection arise, those who do not leave physical progeny behind are not excluded from the notion of the proletariat. As Plato taught, children can be of the body but also of the mind, the latter—creations, inventions, writings, artistic productions...—in fact preferable and closer to the realm of ideas than the former.)

A planetary democracy would still unfold within broken outlines, the singular places and futures of its participants never gathered into a unified whole. But these disjunctions, to be negotiated against the backdrops and the timelines of global environmental degradation, are more in tune with planetary reality than the four- or five-year terms of actually existing national democracies. After all, here, on the planetary level, Trump—not only the empirical person bearing that name but everything that he stands for—would be a life-form among countless others, not the voice that trumps the right of everyone else to have a future.

PART II

CULTURAL UPHEAVAL

1

ON KNEES AND ELBOWS

On May 25, 2020, George Floyd was killed by a police officer who pressed his knee into Floyd's neck for nearly nine minutes. For several years prior to this terrible incident, a growing group of American athletes had been "taking the knee" during the playing of the national anthem at sporting events as a sign of protest against the systemic nature of racism in the United States. The concentration of these entirely contradictory behaviors in the same joint prompted the following reflection.

I have always shied away from thinking and writing about "the body," a theme that had been fashionable in philosophy and social theory at least since the 1980s. Not that I find the topic excessively intimate; on the contrary, it seems to me that *the* body is too idealist, too abstract even, to say anything meaningful about. As a concept, it is still marked by its corralling in the infamous mind-body split that poisons everything on either side of the divide.

Things are different with respect to specific organs. While awaiting my turn at the rheumatology unit of a major hospital, it occurred to me that, holding the bones together and supporting movement, joints are the body parts that epitomize our epoch. Even the complex apparatus of the nervous system does not do justice to our age of connectivity and disarticulation, of linkages and dislocations, to the same extent as joints that also play a literal role in today's politics and everyday life.

Two events have hurled elbows and knees, among other joints of the extremities, into the spotlight: the COVID-19 pandemic and what is now officially confirmed as the homicide of George Floyd in police custody.

During the pandemic, elbows have been entrusted with the work of the hand, whether constructive or destructive, intentional or not. To minimize the possibility of passing the new coronavirus to others, we now take part in an equally new social and respiratory etiquette. An alternative way of greeting that has replaced handshakes is an elbow bump, while coughs and sneezes must be contained in an arm folded at the elbow. A peaceful salutation and danger converge on the same joint (with the minor nuance that the first activates the outer side of the elbow, and the second—the inner pit) and even on the same position of the extremity it is a part of.

The death of George Floyd followed a nearly nine-minute period when police officer Derek Chauvin was kneeling, pressing a side of Floyd's neck down with his knee. In the subsequent widespread protests against police brutality all over the United States, numerous officers, including New York Police Department (NYPD) chief Terence Monahan, "took the knee" alongside the protesters. The practice of taking the knee was, of course, pioneered by American athletes during the playing of the national anthem at sporting events in order to highlight racial injustice in the country. Again, the worst and the best are aggregated, joining together and revolving around actions centered on the knee joint.

Rather than merely ensure the flexion and extension of limbs, granting them the freedom of movement within the anatomical limits of bones, muscles, and ligaments, elbows and knees take charge of acting. That action is immediately combined with a counteraction: every articulation is, simultaneously, a disarticulation. More importantly, the actions and counteractions of the joints write another chapter in the story of a being we still call "human."

From an evolutionary perspective, the upright bipedal posture liberated the hands of our ancestors in a self-perpetuating cycle of humanization. The early humans were humanized by the work of their hands, which, in turn, was made possible through the development of bipedalism. Against this background, kneeling is a symbolic act of consciously giving up the physical (and the associated moral) uprightness of the human stance. It can signal the humility of coming back down to earth or descent into the whirlpools of violence and raw force with its wordless proclamation of the right of the strongest. It points toward humanity beyond and below the human. Similarly, the actions of elbows free the very hands that had been freed for activity thanks to bipedalism. The redoubled freedom of the hands is the crux of the controversial post-human condition, where incredible technological advances coincide with the pronounced loss of fine motor skills and finger dexterity in children.

My point is that the involvement of knees and elbows in some of the recent events is not at all accidental. Our "time out of joint," in the words of Shakespeare's Hamlet, is the time of the joint. But, if the focus of action shifts to knees and elbows, so should the locus of thinking. Instead of thinking with our hands in an attempt to integrate theory and practice and instead of thinking on our feet, we should start thinking on our knees and elbows, with our elbows and knees.

The task is as inherently contradictory as its anatomical support structures: joints are not seamless connections between bones, but ruptured articulations, integrating the skeleton into a whole and allowing for the freedom of movement. To function properly, joints must be out of joint. Are their effects not reminiscent of solidarity in isolation, which many of us have experienced during the pandemic, or of solitude in the midst of social media technologies and the apparently all-embracing virtual networks?

The thinking of the joints and jointures is contradictory because it belongs in the space in-between (bones, and not only). Curiously, the philosopher of contradiction *par excellence*, G.W.F. Hegel does not discuss joints in his *Philosophy of Nature*, where he considers in excruciating detail other aspects of "the animal organism." Joints, after all, do not form a system, which is probably why Hegel omits them at the expense of the skeletal structure. Unless he silently discerns in them the living architecture of dialectics as such.

Now is the moment to invert the habitual scheme of the system and to view time from the standpoint of its ruptures, not continuity, and the bones from the perspective of the synovial joints: the humerus, the radius, and the ulna from the perspective of the elbow; the femur and the tibia from the perspective of the knee. Doing so will give us an embodied approach to the terms that perform relational work, instead of prompting us to concentrate on those that are interrelated through them. As we zero in on the relation itself, we will be privy to a whole slew of upheavals and reversals, the surprising twists and turns that populate the in-between.

2

BEING IN EXILE FROM ONESELF

These reflections date from Yom Kippur, September 23, 2015.

I found myself, this Yom Kippur, reflecting on the exilic condition. With the rift between the State of Israel and the Jews of the diaspora growing wider by the day, could it be that the notion of exile holds the key to understanding what is at stake in the contemporary bifurcated situation of world Jewry?

There is a world of difference between the Latin *exilium* and the Hebrew *galut*. The former is, unsurprisingly, negative; its etymology refers to being away (*ex-*), out of one's place, and straying or wandering there. The latter, on the other hand, is derived from the verb *legalot*, meaning "to reveal," "to expose," "to uncover." Its sense brims with positive connotations, even if the exposure in question may refer to vulnerability, being in harm's way, feeling endangered outside one's native land.

Bemoaned for large stretches of Jewish history, *galut* is, at the same time, the prerequisite for truth and for ethics. Without exposure to the world, particularly outside one's parochial dwelling place, there is neither a meaningful experience nor significant knowledge. Before any verification procedures, truth is what is revealed, from the outside, to our senses and ideas thanks to our exposure to exteriority. Assuming that this exposure has the dimension of precariousness, it entails an ethical demand to respond to the needs and plight of those who suffer the most. Before any moral imperatives, ethics is the way we are called to respond to the suffering of others, or to our own.

Disparaging *galut* is turning a blind eye to the human condition of vulnerability, openness (or openedness) to the world and to the other. The Jews' gathering from the diaspora, the return from exile glorified in the Israeli ideology and in some strands of millenarian Judaism, connotes a closure of the mind and the body, shielding and protecting a phantasmatic identity from exteriority—above all, from the needs and the plight of others. Whereas *galut* is exposure and revelation, "the return" is hiding from truth and from ethical demands, retreating into a shell of provincial belonging, hardening oneself to the world, becoming deaf to suffering, notably of the Palestinian other. Exiled from others, exiled from exile, one enters a strange state of *galut* from oneself, celebrated in everything that is associated with the image of *sabra* (*tzabar*), the newly nativized Israeli.

In his writings, Jewish Hellenistic philosopher Philo of Alexandria (25 BCE–50 CE) implicitly distinguished spiritual exiles from physical ones. The "true *galut*" for him was not "the political exclusion" from ancestral lands but, rather, "the enforced exclusion" from the prophetic mission. Clearly, Philo-the-Platonist felt that the exile of spirit was worse than that of the body. With regard to present-day Zionism, we witness a mutation in this logic: spiritual exile worsens when physical *galut* is supposedly overcome. The bodily occupation of the land, from which our ancestors had been exiled, goes hand-in-hand with the oppression of Palestinian people, who have inhabited this very land. What this disjunction itself uncovers are the difficulties of dealing with *galut*, that is to say, of facing up to ourselves through the attention we pay to the other.

3

THE MUSLIM "NO"

This short essay was first penned in the spring of 2015.

Each of the three monotheistic religions, commonly referred to as "Abrahamic," has its own affirmation of faith, a single statement held to be fundamental by its adherents.

In Judaism, such a proclamation is *Shema* (Listen), drawn from Deuteronomy 6:4. It reads: "Listen, O Israel: The Lord is our God, the Lord is One!" Observant Jews must recite *Shema* daily—for instance, before falling asleep—and it is supposed to be the last thing they utter before dying. Even in the most private nocturnal moments and on one's deathbed, *Shema* announces monotheistic creed, in the imperative ("Listen!"), to the religious community, united around "our God" who is "One."

Christianity, too, has its dogma, which goes back to the Apostles' Creed, dating to the year 150. Still read during the baptismal ritual, the statement of faith begins with the Latin word *credo*, "I believe," and continues, "in the all-powerful God the Father, Creator of heavens and earth, and in Jesus Christ, His only Son, our Lord, conceived by the Holy Spirit, born of the Virgin Mary." The pronouncement individualizes the believer; not only does it start with a verb in first person singular, but it also crafts her or his identity through this very affirmation. While the Judaic *Shema* forges a community through a direct appeal to others, the Christian profession of faith self-referentially produces the individual subject of that faith.

The declaration of Islamic creed is called *Shahada*, "Testimony." In contrast to its other monotheistic counterparts, however, it commences with a negation. Its first word is "no," *lā*: "There is no God [*lā ilāha*] but Allah, and Muhammad is his messenger."

Formulated in the early part of the eighth century, *Shahada* plays an integral part in the conversion process and is the first of the Five Pillars of Islam. The first portion of the "Testimony" is a confession of *tawhid*, or the oneness of God. Its rigorous monotheism hinges on repudiating the existence of any other gods, which, itself, borders on atheism. (The four opening words in the English translation of *Shahada*, "there is no God," may be easily conflated with an atheist conviction.)

Generally speaking, it is striking that the Islamic affirmation of faith is a negation of other deities and religions. Some will, no doubt, take this as evidence of the intolerance lodged at the very heart of Islam. For my part, I do not think things are that straightforward. After all, the Córdoba Caliphate (929–1031) was respectful of ethnic and religious diversity under Muslim rule in what is now Southern Spain. In the world of medieval Islam, astronomy, mathematics, and medicine were thriving. Arabic translations of and commentaries on Aristotle proved indispensable to the recovery and transmission of the Greek classics they helped reintroduce in Europe. So, the question is: how can the same principle of *Shahada* stand behind these developments and the current rise of the Islamic State?

I suggest that the negation, with which the Testimony begins, is the common element motivating the great achievements of Islamic science and philosophy, on the one hand, and the fundamentalist purges of nonbelievers, on the other. The negative form of *Shahada* broaches the indeterminate space of *freedom*, unattached to a specific ethnic community as much as to the subjective identity of the believer. Sweeping the ground clean of all idols, fetishes, and gods, the most recent of the three monotheisms endows its followers with the possibility either to create something new in this clearing or to carry the destructive drive through to its conclusion, destroying and negating itself. There is nothing inherent in Islam as such that could influence the choice one way or another. What proves to be decisive is the historical and geopolitical conjuncture at any given moment, as well as the individual and collective capacity to endure and sustain the heavy burden of freedom.

Amidst the crumbling traditional values of the West, with its own "death of God" announced by Friedrich Nietzsche, the religious "no" waxes more destructive than ever. Its response to passive secular nihilism resulting in apathy, relativism, and the loss of meaning is the active nihilism of fanatical fervor, intolerance, and insistence on the absolute truth… of nothing in the form of the negation. Although it appears that the fundamentalist option is the exact opposite of the liberal West, the two nihilisms (active and passive) resonate with and reinforce each other logically, ideologically, and militarily. Disenfranchised and disenchanted young people from Europe who, having converted to Islam or come from an Islamic background, join the ranks of the Islamic State fail to realize this secret complicity. Adrift and in search for meaning, they fall into the trap of yet another, more deadly, nihilism, which they mistake for a certain and secure foundation lacking in the milieu they are familiar with.

All this is not to say that basic religious pronouncements in the affirmative, like the Judaic *Shema* or the Christian *Credo*, are in any way superior to the basic negation in *Shahada*. Quite the contrary: the inaugural "no" holds a greater potential for freedom than they do. Nor do I claim that every Muslim person and community responds to the provocation of negativity in the same manner. Indeed, many in the past and in the present have embarked on a more difficult path of radical enlightenment and creativity, indebted to the dismantling power of the negation. But as the battle for hegemony in the Muslim world rages, it is crucial to understand what is at stake in the most recent incarnation of fundamentalist destructive fury, where it is situated on the global theologico-ideological map, and what the alternatives to the thoughtless dismissals (or endorsements) of Islam, so prevalent today, are.

4

DON'T KEEP CALM! AND DON'T CARRY ON!

In 2014, I inquired into the ideological function of the proliferating calls to "keep calm and carry on" amidst the unraveling status quo.

For some time now, there has been a tremendous proliferation of T-shirts, mugs, signs, bags, and other merchandise urging us to "keep calm." More than that, there are now countless variations on the "Keep calm and carry on" theme, initially invented by the British Ministry of Information at the outset of World War II.

These posters, never officially issued and largely forgotten until the start of the new millennium, were meant to reassure the British public, wary of possible air strikes on cities and towns. Today, everywhere we encounter ironic and trivializing spoofs on the original propaganda and morale-boosting campaign, such as "Keep calm and marry Harry," coined on the occasion of the wedding of Prince William and Kate Middleton; "Keep calm and drink tea"; or, to stay on the topic of nutrition, "Keep calm and have a cupcake", strangely reminiscent of Marie-Antoinette's infelicitous advice to the starving masses deprived of bread, "Let them eat cake!" Although the message is still very much associated with the United Kingdom (if not with Britishness, down to the monochrome background, the font and the royal insignia that usually accompany it), its visual presence and ubiquitous effects are now undeniably global. If, from so many corners, we hear echoes of the 1939 call, this is because the danger is real—that the public is about to lose its collective cool. For, one of the intense effects that surfaced as a result of the Euro crisis was, precisely, public anger at the growing misery, hopelessness and un- or under-employment for the vast majority of the increasingly impoverished European population.

In Spain, the sense of indignation was strong enough to lend a name to the popular protest movement of the *Indignados*. A corollary to this emotional state, shared by many, was the prerevolutionary feeling that things could no longer go on the way they had been and that drastic political and economic changes, perhaps amounting to the end of capitalism, were imminent.

In contrast to the boiling sentiment of public anger, "Keep calm and carry on" is the expression of ideology at its purest. The seemingly permanent economic state of emergency we are living through resonates with the political state of emergency, wherein the poster historically originated.

The austerity programs, implemented all over Southern Europe, amount to one way of "carrying on", sanctioned by national conservative governments in concert with the infamous Troika (the European Commission, the European Central Bank, and the International Monetary Fund). Rather than take the crisis as an opportunity for a radical restructuring of the European Union, giving it a more explicitly political, integrated, and egalitarian character, the bitter medicine the Troika administered was the destruction of the welfare state in Southern Europe and stagnation in salary growth in the rest of the continent. All the while, those suffering the most from such measures were urged to—you've guessed it!—keep calm and await the light at the end of an unending tunnel.

Popular culture is rife with appeals it draws directly from the core desire of mainstream ideology and feeds into music, cinema, and other arts. In the 1980s, a popular song "Don't worry, be happy" provided the soundtrack for the highly destructive neoliberal agenda of Ronald Reagan and Margaret Thatcher, the late US and UK leaders, respectively. "Keep calm and carry on" is admittedly more disenchanted, more sober, and even more stoically resigned than that. There is no promise of either happiness or salvation in it—just the drudgery of ongoing sacrifice, carried out with perfect serenity, for the sake of the sheer continuation of existence.

Now, that is precisely the point: the main achievement of the status quo is to convince those who live under its sway that their (our) lives and futures are contingent on *its* extension into the future and that, without it, they (we) will not survive, not even for a day. Fears of financial collapse and social chaos are sown in order to show how the current, admittedly imperfect, arrangement is better than the alternatives.

At the extreme, losing my cool—our ideologues imply—would result in undermining my capacity to carry on, to go on living. Worse yet, if members of the public do not keep calm and maintain a basic trust in the political-economic system they live under, then their collective existence is threatened, together with this very system. What philosopher Baruch Spinoza called our *conatus essendi*, the love of life that literally ties us to our essence or our being, becomes a powerful political mechanism that, in the same gesture, binds our individual and common destinies to the fate of the powers that be.

To turn the essentially conservative message around, it is necessary, in the first place, to say "no" to it, to negate it with the appeal: lose your cool, stop going about your life as usual! To a certain extent, the economic and environmental crises we are mired in have this disquieting effect on us, the effect that the ideological constructions of normalcy try to assuage. Cutting through layers upon layers of obfuscation, spawned by these constructions, we finally realize a deeper meaning of the contemporary crises, namely, that our only chance at carrying on and surviving hinges on our refusal to keep calm and on untying our destinies from that of the highly destructive status quo.

5

UNCULTURED AUSTERITY

Austerity policies in Southern Europe had a tremendous impact on the public funding of culture in the beginning of 2010s. The reflections below date from 2012.

Culture is the silent victim of the Euro crisis. Last month, the Portuguese government announced the definitive closure of 38 cultural and social foundations and 100 percent cuts in the funding of 14 more. Likewise, the Spanish government reduced public funds allocated to cultural organizations by 70 percent in the past three years. Despite State Secretary of Culture José María Lassalle's previous affirmation that culture is "neither a luxury nor a caprice," the new budget of Prime Minister Mariano Rajoy's government did not spare the hallmark Prado Museum, the Institute of Cinematography, or even the Network of Public Libraries, which will receive no money for new books next year. Meanwhile, in the Netherlands, arts funding has been slashed by 25 percent. And Italy's La Scala opera house faces a $9 million shortfall, owing to reductions in subsidies.

Such decisions rest on the premise that culture is nothing but superfluous entertainment. After all, who needs circuses when there is no bread? But this assumption is deeply flawed. Strategies for overcoming the Euro crisis have been rooted in the same narrow-minded economic thinking that led to it in the first instance. As a result, fiscal deficit continues to grow, while strict austerity programs designed to reduce it have merely undermined business and consumer confidence and damaged people's well-being.

Cultural production, the wellspring of collective imagination, generates resources for finding creative solutions to problems, even those that seem completely unrelated to art. In fact, on the implacable terms of the market, cultural production—one of those rare "industries" that has shown persistent, long-term growth over the last decade—could well be a much-needed niche for

European exports. Policymakers stubbornly continue to overlook this potential. For example, in Cascais, Portugal, Casa das Histórias, a museum dedicated to the artist Paula Rego, has been a great success, with 300,000 visitors since its opening in September 2009. Nevertheless, it is slated to close down, the more than 500 works that Rego donated to be deposited in the Cascais city hall.

In her work, Rego explores humanity's unconscious, animalistic underside. For example, through her depictions of women and children as flesh-and-blood beings, rather than as idealized figures, she challenges viewers to confront what they usually ignore, the uncanny, uncomfortable small verities of the human condition. Casa das Histórias, like other museums and cultural organizations, provides visitors with an opportunity for self-reflection, which is much needed in times of crisis.

Cascais's ex-mayor, António Capucho, called the decision to close Casa das Histórias "barbaric," saying that it reflected an "error in calculations." But the recent announcement of closure by the Portuguese government showed that the error was never corrected. After all, when austerity undermines art, culture, and imagination, it defies critical thinking. To overcome the crisis, policymakers must reject solutions born out of old modes of thinking, and begin to think anew. Sustaining and fostering cultural production is a good place to start.

6

A GENEALOGY OF ENJOYMENT

A short study from 2012, speculating on the theological framing of enjoyment.

In a hedonistic culture, obsessed with the maximization of individual pleasures, it is easy to forget that "enjoyment" was once a rigorously theological concept with strong ethical overtones. The reminder comes from unlikely quarters: Hannah Arendt's 1929 dissertation on *Love and Saint Augustine*. There, Arendt highlights a fundamental distinction in Augustinian thought between enjoyment (*frui*) and use (*uti*), attitudes appropriate to human relations to God and to the world, respectively. Strange as it may sound, according to St. Augustine's *Christian Doctrine*, to love God is to enjoy Him (*Deo frui*), which is to say, to relate to Him as an end in itself, not a means for external ends. Centuries later, Immanuel Kant will transpose this love from divinity onto other human beings, substituting the language of respect for the discourse of enjoyment. An austere ethics of reason germinates on the ruins of the Augustinian distinction.

In contrast to Kant, proponents of the utilitarian "calculus of pleasure" seem to have preserved the ethical significance of enjoyment in a thoroughly secular context. Didn't Jeremy Bentham's principle of "the greatest happiness for the greatest number" democratize the divine prerogative? Answering in the affirmative, we would overlook a significant nuance of the original Augustinian distinction, namely that to enjoy was to refrain from usage, to abstain from all considerations of utility. Nothing in the world, including our own bodies, was exempt from the possibility of being used for human purposes. Only the Supreme Being or the Highest Good, for the sake of which human beings, too, existed, merited the attitude of use-less enjoyment. All other creatures were subject to rampant instrumentalization, implementing the divinely sanctioned transformation of the entire world, the mundane realm as such, into a collection of raw materials for present and future consumption.

Despite these serious problems, St. Augustine's writings hold out a beautiful promise for all those who wish to work out the logic of enjoyment without use. In Kantian and post-Kantian modernity, a pale afterglow of this promise typically lights up in the aesthetic experience and the disinterested pleasure proper to it, insulated from theoretical explanations and concerns with utility alike. But even art failed to save enjoyment in the robust sense, which had, by the time of Kant, long deteriorated into "mere" enjoyment, the entertainment value making artworks agreeable and so detracting from the disinterest with which we ought to approach them. More gravely still, in the aftermath of its thorough aestheticization, enjoyment no longer stood for a principle of ethical action—the position it had occupied at least since the Antiquity of Epicurus—but, instead, became passive and utterly inefficacious. In a word, art was a far cry from a solution.

An alternative, if not a utopian, path lying before us is one St. Augustine himself considered both heretical and insecure: to enjoy the world elevated to the now-vacant place of God (and of the human). Such elevation would not extinguish the possibility of utilizing certain entities in the world, for instance, as nutritional sources; it would, rather, require a drastic shift in mindsets away from the appropriative notion of who we, ourselves, are and promote an understanding that consumption does not constitute the default mode of our relation to the world. Differently stated, use would become an exception from the general rule of ethical enjoyment.

The Latin verb *fruor*, to enjoy, at the heart of the human relation to God in St. Augustine, is drawn from the natural world, or, more specifically, from the world of vegetation. It is not an idle curiosity that "enjoyment" shares its grammatical root with "fruit." If etymology were to be taken as evidence, we could conclude that fruits are the prototypes of unlimited enjoyment by whoever cares to pick them, that is, that they are there *for* the eaters' pleasure. Nonetheless, in light of St. Augustine's interpretation of *frui* in opposition to *uti*, fruit, like the plant that bears it, must also be an end in itself, a carrier of intrinsic value, a vehicle for the perpetuation of the plant, an incarnation of what permits it to share in the immortal through reproduction, as Plato has it. Just as the fruit stands at the intersection of usage and use-less enjoyment, so everything else in the world may fall into one or the other category, depending on the kind of comportment humans adopt toward it.

In the Augustinian universe, the love of the world was misdirected and near-sighted, at best. It represented a diversion from the love of God, a dispersion of enjoyment in numerous insecurely possessed things, and, hence, a renunciation of true Being. The repudiation of finitude made sense in the context of faith in the existence of perfection above this world, the perfection against which the world was measured (failing, of course, to measure up to it) and to which it could be sacrificed as a whole in the hope of salvation.

Once the transcendent absolute becomes questionable, the certainty that the world exists for the sake of something outside itself also fizzles out. That St. Augustine's distinction has not yet caught up with this momentous event is evident in the continuing treatment of the world as though it were a collection of utilizable objects, in the absence of anything meriting our use-less enjoyment, save for works of art. What we urgently need to learn in the process of mourning the loss of absolutes is the sense of a modified Augustinian motto, *Mundo frui!* Enjoy the world!

7

THE TWO SUNS OF EUROPE

> This 2013 text presents a set of prolegomena to an alternative political heliocentrism.

It all revolves around the sun. Or, better, around the suns.

The image of Europe is divided along the following lines. On the one hand, there are the sun-soaked streets of Madrid, Lisbon, Athens, and Rome; on the other hand, there are the gloomy avenues of Berlin, London, and Amsterdam, illuminated by another kind of light—that of knowledge and of the Enlightenment. Southern Europe, in the eyes of North Europeans, is a place of leisure and laziness. It is sold (and bought) as a convenient vacation destination, where the Northerners can partake of a siesta-like lifestyle for two weeks a year or less. On the beaches of Algarve or on the Greek islands, they can forget their worries and temporarily lose their "bearish seriousness," as Nietzsche liked to call it. Still, Southern Europe is not exactly the Homeric Island of Circe, where visitors forgot their native lands. As soon as they get a little tanned, or sometimes thoroughly burnt due to their inexperience, our contemporary Odysseuses will go back home to their routine, reenergized for yet another year of work and solemn dedication to all sorts of rational endeavors.

This caricature has more than one grain of truth, mixed with the grains of sand and salt that get caught in the hair of sunbathers on the beaches of Ibiza or Alicante. It is, to be sure, not a truthful representation of Southern Europe but of the typical attitude toward it in other corners of the continent.

The severity of the current economic crisis in Spain, Portugal, Greece, Cyprus, and Italy is overviewed through a uniform ideological lens. The populations of "peripheral" countries are lazy, carefree, refuse to work hard, and instead bask in the warmth of the actual sun, which they enjoy in abundance. They have lived long enough on the money of the German taxpayer—the average tirade continues—and should now suffer the consequences of their economic shortsightedness.

It matters little to those who hold such a view that the Portuguese, for example, must endure much longer working hours than the Germans, or that South European countries have provided easily accessible consumer markets for goods made in the heart of the European Union. It matters even less that, in the European context, the sun of philosophy first rose in Greece before migrating to Rome. The prejudice is, by now, quite fixed: the physical sun shines in the South, whilst that of knowledge illuminates the North.

We cannot neglect to mention the philosophical backstory for the two suns of Europe. In his *Philosophy of Nature*, Hegel goes so far as to compare the psychology of the Southerners to that of plants, which also strive toward the light and warmth of the sun. "The externality of the subjective, selflike unity of the plant," he writes, "is objective in its relation to light. [...]Man fashions himself in more interior fashion, although in southern latitudes he, too, does not reach the stage where his self, his freedom, is objectively guaranteed."[1]

Translated into ordinary language, Hegel's statement means that plants are driven by something outside of themselves, namely light, from which they derive their identity. Humans, on the other hand, build themselves up from within, as conscious and self-conscious beings with memory, decision-making capabilities, and so on. *But*—here comes Hegel's racist punch line—human subjectivity in "southern latitudes" is more plant-like and the human self is neither as free nor as fully developed as that in the northern latitudes. Although he often opposes Hegel, Nietzsche would agree that the overabundance of light does not leave enough time for gloomy rumination, dwelling on the deep and dark recesses of the soul (and this lack of time is a good thing, he would add, laughingly). The amount of light from the metaphysical sun of knowledge and the intensity of the physical sun are inversely proportional: the more we receive of the latter, the less we benefit from the former.

Of course, tourism is a crucial industry in Southern Europe, which is why it has been only all too eager to confirm its status of the "sun destination." Spain's 1982 advertisement campaign with the logo designed by Joan Miró and the motto "Everything under the sun" (*Todo bajo el sol*) has cemented the common association of the entire country with beaches and entertainment. Portugal's more recent program *Reforma ao sol*—"Retirement under the sun"—has aimed to attract North European pensioners (and especially their pension funds) to an environment propitious to relaxation after a lifetime of work.

So long as the different zones of Europe remain lit by two different suns, the talk of "two-track Europe" will be on the political agenda. It is not within the realm of the possible to prevent rain from falling over Benelux countries or the British Isles. But it is within our power to develop knowledge-based economies in Southern Europe. Until that moment arrives, European "integration" will be an empty word.

8

FOR THE LOVE OF A CITY

> Composed in the last days of 2012, this essay muses about a specific form of contemporary affect, the love of a city.

People confess their love for cities all the time, everywhere: on T-shirts and caps, bumper stickers and water bottles. "I ♥ NY" is probably one of the most successful marketing campaigns geared toward promoting tourism ever.

But what does it mean to love a city? Do the lovers of a contemporary megalopolis, such as New York, feel affection for every one of its millions of inhabitants? For all the buildings, traffic-clogged streets, highways, and sewage systems that make up its infrastructure? Do they love the Big Apple as a whole? The seemingly infinite possibilities for entertainment it offers? The feeling of being in the right place at the right time: Time Square, New Year's Eve?

When we shower our love onto a city, it is most likely an idealized image that we hold so dear. The shabby reality of everyday urban life is either overlooked or not registered altogether: few tourists visit the Bronx or Queens, those parallel worlds that are practically set apart from the glamor of (certain areas in) Manhattan. To confess our love for a city is to conjure up a utopia, an object of affection that does not really exist, and to transpose it onto an actually existing place.

That is how easily recognizable landmarks—Empire State Building, Brooklyn Bridge, and so on—acquire their larger-than-life significance. These privileged parts of the city stand in for the whole in a substitution that makes the process of idealization possible. New York "is" Chrysler Building, and it "was" the Twin Towers. The magical feeling of being there is, in part, due to the enchantment of these parts that miraculously condense the whole, while hiding from sight the city's messier facets. One knows that one is in New York so long as one catches a glimpse of the imposing rooftops of Empire State and Chrysler Buildings. The rest does not matter.

For those who "♥ NY," the classical Christian distinction between the heavenly and the earthly cities no longer applies. Sections of the megalopolis stretching here below are invested with the function of representing a secular heavenly ideal, which is coincidentally not all that different from what St. Augustine and others would have deemed to be "sin city." There, all our wishes come true, as fiction and reality blend in a Hollywood-infused haze.

I wonder, however, if a different love of a city is conceivable. There is no doubt that the inhabitants do not appreciate their hometown in the same way as tourists do. But even then, many pass an entire lifetime without ever setting foot in some of their own city's stigmatized neighborhoods.

So, if living in a city is not sufficient for loving it otherwise, then what is? Nothing less than a concerted and conscious resistance in the face of a beclouding vision of the city as a picture-perfect object. Only if you are not discouraged by direct contact with the dark undersides of the urban jungle, will you earn the honor of tough city love.

The hundreds of thousands who will gather in New York's Time Square to watch the ball drop on December 31, 2012, will be drawn there by their love of the city with its promise of being at the center of the world. The few who will linger after the festivities are over will observe confetti- and garbage-filled streets, frantically cleaned by crews of municipal workers, for whom the New Year's is just another day on the job. Will these onlookers have the heart to proclaim, at that very moment, "I ♥ NY"?

9

WHAT HORSE MEAT TELLS US ABOUT OURSELVES

In the 2013 horse meat scandal, tests of ground beef in various EU countries discovered high percentage of undeclared horse meat in this product, sometimes amounting to 100 percent.

All of us have heard the age-old adage, "I am what I eat." As late as the nineteenth century, Friedrich Nietzsche took this dictum literally and turned it into a cornerstone of his physiological psychology. National diets, in his half-playful reconstruction, gave rise to distinct patterns of digestion (or, in many cases, indigestion), which accounted for various personality and cultural types: the ruminative-contemplative or the spontaneously active. Spirituality and the richness of inner psychic life were correlated to a slower metabolism and inaction. More crudely put, the ethereal realm of spirit was born of the accumulation of gas in the intestines and the inability to pass it otherwise.

Today, over one hundred years after Nietzsche's death, we have very little idea of what we actually eat. The EU meat scandal should prompt robust discussions of dietary ethics and of consumer awareness when it comes to the ingredients of industrially prepared foodstuffs; the shocking proportions of horse meat in ground beef are but a reminder of our ignorance about the contents of our meals. And that is not to mention preservatives, bleaching agents, sulfites, and other chemical ingredients that are even more common in food items than salt and sugar, with which they are laced in a vast majority of cases.

But, aside from the health concerns we all share, what interests me as a philosopher is a simple syllogism. If (1) we are what we eat, and (2) we do not exactly know what we eat, then it necessarily follows that (3) we do not know who we are. "Know thyself!"—another tidbit of ancient wisdom—begins with knowing what is on your plate.

Much has been made in contemporary thought of the indeterminacy of human beings. We cannot be defined in our essence, in terms of who or what we really are, except as pure possibilities, capable of being virtually anything. Could it be the case that the sophisticated shoulder shrug in response to the question about the meaning of being human is due to the indeterminacy of our diets, at least from the standpoint of the consumers themselves? Will we become something other than pure possibilities once we know precisely what it is that we are gulping down at any given moment?

All jokes aside, the fullest conscious knowledge of the ingredients that go into our diets is not sufficient to dissipate human indeterminacy. The absorption of nutrients passes below the radar screens of our mind, as do other processes comprising our metabolism. Some murkiness will always remain, testifying to the non-transparency of the mind's relation to the body. But the most crudely material, if ever so incomplete, knowledge of ourselves will invariably have to begin there.

Public outcry over mislabeled meat in the European Union was not primarily motivated by health concerns, even though some of the horse meat mixed into ground beef was probably tainted with veterinary drugs harmful to humans. Instead, people felt betrayed by the food industry that created an extra layer of opacity between them and what they ate, that is to say, between them and… themselves. A tainted frozen lasagna dinner has become a symbol of something other than a cheap meal for the growing number of the working poor and those affected by the ongoing economic crisis. Henceforth, every time we look at food items displayed on a supermarket counter, we will imagine how they are laughing at us: "You do not know the least thing about yourself, not even what it is that you put in your mouth!"

10

CONTAGION: BEFORE AND AFTER COVID-19

The COVID-19 pandemic prompted me to take a closer look at the notion of contagion.

With the new coronavirus pandemic, the sense of touch has come under attack. Medical authorities insistently advise us: do not touch your faces, do not touch doorknobs with your hands, and definitely do not touch others (no kisses on the cheek, handshakes, or other bodily greetings). It is easy to understand the epidemiological reasoning behind such guidance. The virus is highly contagious and can survive not only on the skin but also on inorganic surfaces for a relatively long period of time. But the cultural frame of what is going on is at least as important as the biological and epidemiological explanations.

The acceleration of processes that went under a broad heading of globalization involved a growing virtualization of the world. To become a globe (which is, itself, a geometrical abstraction, rather than lived reality), the world had to be reimagined as an ideal unity. Cultural homogenization, often opposed and criticized for impoverishing and razing local customs and ways of life, was but a side-effect of that idealizing movement.

A lion's share of the work of virtualization was accomplished by the Internet, along with cutting-edge technologies for obviating or digitalizing touch: voice-controlled devices, touchless water taps, soap dispensers and flushes, and contactless payment methods. Touch became obsessively linked to the touchscreens of smartphones and tablets that, as we have now learnt, could be perfect sites for spreading the infection.

In a nutshell, it was a dream of globalization to create a touchless reality, virtual togetherness without bodily involvement. Tourists and politicians, UN workers and expatriate professionals, business consultants and (yes!) academics could move throughout the globe as though they were disembodied spirits, present in the flesh as if they were not really present, not really exposed to sudden dangers and potentially lethal contingencies.

In the age of the new coronavirus, that dream has been shattered. It turns out that our means of transport carry more than their human passengers and that locations, which could just as well be anywhere, like airports or big hotel chains, also host other forms of life or nonlife—the viruses. This uninvited and invisible excess reminds us of the fact that, far from disembodied spirits, we exist thanks to our fragile bodies that may, in a matter of days, find themselves on the verge of serious illness or death.

The COVID-19 pandemic signals a return of what has been suppressed or repressed by the entwined drives to globalize and virtualize the world, namely the body. And the body is represented by the sense of touch, its corresponding organ being not the hand, as we automatically assume, but skin. Just as jetlag testifies to discrepancies between our technological and physiological temporalities and possibilities, so the current pandemic highlights the incongruities between the dream of virtualization-globalization, on the one hand, and living "in one's skin," rather than purely "in one's head," on the other.

Here, a seemingly negligible difference between infection and contagion comes to the fore. Infection means being tainted within, receiving something impure inside oneself. Contagion, as a Latinate rendition of "with-touch" or "touching-with," entails not the infiltration of a foreign and potentially damaging substance into an organism, but communicability, the ability of pathological agents to jump from host to host (as in "communicable disease"). Like the touch it names, contagion happens on the surface, at the interface of surfaces in contact with one another. It leaves the fiction of contamination-as-infiltration behind and, instead of fixating on the distinction between the inside and the outside (and on the boundaries, borders or membranes separating the two), focuses on the territory of contact, the space inbetween that makes every body but a transit station for viral self-replication.

The other peculiarity of contagion is that it disrespects divisions between areas we typically treat as totally independent: biology and economics, psychology and informatics. Viruses replicate themselves in living tissues and computer programs, through memes that "go viral" or through the DNA or RNA encoding of proteins, by which they act. Contagion spreads among members of a population, from one species to another (as in the case of coronaviruses), in financial markets, through rumors and fear, or in the dissemination of ideologies, even as it produces multiple feedback loops between these different areas.

Why do contagions have a significantly broader range and reach than infections? Is it not because contagion requires no more than the brushing of surfaces: of skin and skin, skin and a doorknob, fear-laced gossip and an ear receptive to it, an insolvent bank and similar institutions that lose investor confidence, financial markets and direct economies, "America First" and another nationalist "Me first"? Contagion, then, is all about touch, not incorporation—a factor that lends it its speed and the capacity to spread far and wide.

To return to the coronavirus pandemic: although the dynamics of touch call us back to our bodies, they do so under the sign of more severe repression still. If, before the current crisis, we simply forgot the body, failing to notice it as, without friction, it moved through and interacted with other bodies and surfaces of a globalized world, now we recall (better: we are recalled to) its existence in an atmosphere of conscious negation, distilled in the injunction "do not touch," not even yourself. (Does this injunction not parody the words Jesus addresses, according to John 20:17, to Mary Magdalene who recognizes him after his resurrection: *Noli me tangere*, "touch me not"?)

Strangely enough, the moment masses of people around the world are faced with the fragility of their bodies and lives—the moment we are called back to our corporeal existence—we must take measures to reduce direct contact with others, to retreat to the private cocoons of our homes. No sooner does the flesh-and-blood body make its comeback on a global scale than the virtualization of existence intensifies, with social, cultural, and intellectual life migrating almost entirely to the Internet, which made the virtualization of our coexistence possible to begin with. In a pandemic, the dynamics of touch retrieve the body both as dramatically threatened and as a source of threat, reinstating the strictest version of its virtualization.

A period of respite from the fast-paced routine of our lives afforded to many by the COVID-19 pandemic should be an occasion for reflecting on what was going on before this viral outbreak and what the world, our relations to each other and to our own bodies might look like after it subsides. How does the virus intervene into the history of touch, shaped by social conventions, political and medical regimes, technological inventions? What are the senses of contagion spanning the increasingly "touchless" virtual reality and its suppressed underside? How to live and to think on the surface, with the overlapping (potentially contagious) surfaces we have mistaken for discrete things: bodies, economies, information systems, and systems of beliefs?

PART III

INTELLECTUAL UPHEAVAL

1

A FIGHT FOR THE RIGHT TO READ HEIDEGGER

The fallout from the publication of Martin Heidegger's philosophical diaries, known as *Black Notebooks*, was swift and had very wide repercussions. In spring 2014, calls to ban texts by Heidegger and some other philosophers were sounded in the academic quarters.

This spring, the Students' Union at University College London banned meetings of a group called the Nietzsche Club, which was formed to discuss the ideas of philosophers who inspired, among others, far-right politicians and leaders of the past, like Benito Mussolini, an admirer of Nietzsche's work. The Union's council decided that the discussion of such thinkers and ideas would foster a dangerous wave of fascism among its students and prevented them from holding a public meeting.

To those of us in philosophy concerned with ideological censorship, this incident is but the tip of an iceberg of an impending struggle over the prospects of a serious scholarly engagement with some of the most important philosophers of the nineteenth and twentieth centuries. But, unlike the actual Arctic ice sheets that are melting at an alarming rate, the freeze imposed on thinking is showing no signs of abating. In particular, there is a menacing chill forming around the work of Martin Heidegger.

With the publication of volumes 94–96 in Heidegger's *Complete Works* containing the infamous *Black Notebooks* (or private diaries, not yet translated into English) earlier this year, his critics, pointing at the incontrovertible evidence of Heidegger's anti-Semitism, now claim that his philosophy is suffused with objectionable ideas through and through. So much so that the critique of modernity developed by the German thinker is being reinterpreted as a way to "launder" his anti-Semitism.

As a Jew who suffered from anti-Semitic discrimination in the final years of the Soviet Union, I am weary of the contemporary manifestations of this hateful ideology. But I also find irksome the attempts to use the label *anti-Semitism* as a tool for silencing dissent. Both opposition to Zionism and the thinking inspired by Heidegger now incur this charge, which is leveled too lightly, thoughtlessly, and therefore without a minimum of respect for the victims of ethnic or religious oppression.

Of course, none of the recent revelations about Heidegger should be suppressed or dismissed. But neither should they turn into mantras and formulas, meant to discredit one of the most original philosophical frameworks of the past century. At issue are not only concepts (such as *being-in-the-world*) or methodologies (such as *hermeneutical ontology*) but the ever fresh way of thinking that holds in store countless possibilities not sanctioned by the prevalent techno-scientific rationality, governing much of philosophy within the walls of the academia. It is, in fact, these possibilities that are the true targets of Heidegger's detractors, who are determined to smear the entirety of his thought and work with the double charge of Nazism and anti-Semitism.

Were canonical philosophers to be blacklisted based on their prejudices and political engagements, then there wouldn't be all that many left in the Western tradition. Plato and Aristotle would be out as defenders of slavery and chauvinism; St. Augustine would be expelled for his intolerance toward heretics and "heathens"; Hegel would be banned for his unconditional admiration for Napoleon Bonaparte, in whom he saw "world spirit on horseback." As for Heidegger himself, admit it or not, those minimally versed in his thought will know that his anti-Semitism contradicts both the spirit and the letter of his texts, regardless of the ontological or metaphysical mantle he bestows upon anti-Semitic discourse. It is likely that the German thinker himself did not sense this contradiction, but this does not mean that it was not there. Let me give you an example.

In one deplorable turn of phrase in *Black Notebooks*, Heidegger writes about the "worldlessness" of Judaism and associates the Jews' uprooting from a national territory with the "world-historical 'task' of uprooting all beings from Being," which, according to Heidegger, Judaism presumably shares with modernity as well as with Bolshevism, Americanism, British imperialism, and so on. The French philosopher Emmanuel Faye is correct to trace this concept of "worldlessness" that describes the state of an inanimate object, such as a stone, back to Heidegger's 1929 course on *The Fundamental Concepts of Metaphysics*. As worldless, the Jews are reduced to the level of things—a classical dehumanization technique. But from this valid argument, Faye jumps to a ridiculous conclusion that "the Heideggerian notion of 'being-in-the-world,' which is central to *Being and Time*, may take on the meaning of a discriminatory term with anti-Semitic intent."[1] While his first point probes the depths of Heidegger's anti-Semitism, the second is an amateurish trick, endeavoring to taint a fecund idea by means of free association.

Well before the publication of *Black Notebooks*, Heidegger's organicist metaphors for spiritual life that is rooted, plantlike, in the native soil (for instance in *Discourse on Thinking*) could be read as denying genuine talent and creativity to those who did not enjoy a strong connection to the "home ground," including, in the first instance, the Jewish people. But such racist nearsightedness does not at all follow from the content of Heidegger's philosophy. In response to his error, one could say that the Jewish mode of rootedness was temporal, rather than spatial; before the Zionist project undertook to change this state of affairs, the Jews were grounded only in the tradition, instead of a national territory.

Such grounding is anathema to the uprooted condition of modernity, with which Heidegger hurriedly equated Jewish life and thought, and in which he diagnosed, precisely, the destruction of tradition. From the perspective of the author of *Being and Time*, the temporal nature of Jewish rootedness should have been viewed as more desirable than spatial ties to the soil. After all, didn't Heidegger want to make (finite) time, rather than space, fundamental to human existence?

There is, then, a profound disconnect between Heidegger's anti-Semitic prejudice and his philosophy, which influenced a number of prominent Jewish thinkers, from Hannah Arendt to Jacques Derrida, and from Leo Strauss to Emmanuel Levinas. Yet, more and more, one is forced to justify the very act of reading his works for purposes other than denunciation and censure. As my colleague Marcia Cavalcante Schuback (who translated *Being and Time* into Portuguese) and I write in our commentary on Heidegger's 1934–35 seminar analyzing Hegel's political philosophy: "'The case of Heidegger,' or 'l'affaire Heidegger,' as the French call it, is the case of philosophy facing the loss of its right. And what are all the controversies surrounding Heidegger's Nazism about if not the right of and to his thought, not to mention the right to think further on his path, despite, against, or with his past?"[2]

Formulated more broadly, the question is about who has the right to pursue philosophy, to call herself or himself a philosopher, and to deny this appellation to others. In his book, *Heidegger: The Introduction of Nazism into Philosophy*, when referring to Heidegger, Faye often renders the word philosopher in quotation marks. The current fight for the possibility of reading certain philosophical works is, therefore, a fight over the very meaning of philosophy, with or without quotation marks.

2

HEIDEGGER'S THINKING TODAY IS, PERHAPS, THE POSSIBILITY OF THE WORLD

Brief considerations on the occasion of the publication of my book *Heidegger: Phenomenology, Ecology, Politics* in 2018.

In the 1957 lectures he delivered in Freiburg under the title "Basic Principles of Thinking," Martin Heidegger speculated that "dialectics today is, perhaps... the actuality of the world [*Weltwirklichkeit*]."[1] For all its hyperbolic thrust, one should not take this assertion lightly, dismissing it as a dated intellectual artefact from the Cold War era, when antithetical political camps were locked in a life-and-death struggle on a world scale. Speaking against such an easy historicizing explanation is the fact that the insight cropped up as Heidegger reflected on nothing less than the very foundational principles of thinking. Another piece of evidence corroborating its seriousness is that the notion of the world, presumably actualized by dialectics in a "today" that is sixty-odd years old now, is itself a cornerstone of Heidegger's philosophy. So, what is going on here?

Heidegger's point is that dialectics, be it of the Hegelian variety or the Marxist iteration of dialectical materialism, has long ceased being either an abstract idea or an applied political ideology intended to explain reality in the simplest terms imaginable. Dialectics actively *determines*, commands, and steers the course of the world, split into camps sharing the same general goal: to master, subdue, and appropriate the earth. Fractured and conflictual, the world's dialectical actuality is rooted in a silent consensus of overtly opposing parties, namely that the true purpose of world domination is the seizure of the earth. Far from an opportunistic aberration, this goal inheres at the heart of Western thinking. The ideal capture and appropriation of the object are the means for, and the end of, the real imposition of the thinking will upon whatever and whomever it captures. Dialectics thus accomplishes the mission of thinking with unprecedented effectiveness and success.

Despite simmering new tensions between Russia, on the one hand, and the European Union and the United States, on the other, the Cold War is over. Heidegger's "today" is no longer ours. And yet, it is utterly relevant. Dialectical actuality makes sense within the broader project of constructing a world (frameworks of meaning, extending all the way down to the meaning of meaning) deployed with the view to appropriating and dominating the earth (the ultimately meaningless source of meaning, that upon which life unfolds and into which it returns to commence again) in the shape of territories to conquer or natural resources to extract. The triple knot of phenomenology, ecology, and politics is as tight as ever: a network of lived meanings undergoes behind-the-scenes political integration, or disintegration, such that its elemental substratum is, at the same time, controlled and threatened, secured and rendered fragile, appropriated and pushed to the brink of nonbeing.

It is in this rather depressing light that I would like to update (and so, in some sense, to actualize) Heidegger's assertion for our "today" in the following way: Heidegger's thinking today is, perhaps, the possibility of the world. Immediately, readers will retort that I am indulging in a hyperbole more blatant still than Heidegger's take on Hegel. How can a one-time card-carrying member of the National Socialist party not only gain admission into the philosophical canon but also become pivotal in contemporary thought, not to mention in the contemporary world?

As I argue in my book on the German philosopher, with reference to the contributions of his Russian translator Vladimir Bibikhin, it is a gross mistake to consider Heidegger's thinking as a piece of intellectual private property. In its enduring relevance and generativity, Heidegger's thinking is not his own; it is the thinking of the world. Its lacunae and pernicious blind spots are, to be sure, the thinker's responsibility, chief among them the unquestioned persistence of anti-Semitic prejudices in reflections on the agency and figures of uprooting, displacement, and what we now call globalization. But they are just that—lacunae of the unthought in the midst of the world thinking itself on the hither side of the modern distinction between subjects and objects, theory and practice.

Even then, I raise the stakes of my claim that Heidegger's thinking is, perhaps, the *possibility* of the world today. Given his phenomenological approach to the possible disentangled from its deficient position in a strictly teleological order, existence understood existentially retains inexhaustible possibilities. For the finite world as the domain of existence to be, it must still be possible up to its demise. The possibility of the world as world is, furthermore, exposed the moment it is overshadowed by a grave danger, the moment its time is almost up and it may no longer be possible—say, after a nuclear Armageddon or as a result of catastrophic global climate change. By emphasizing the priority of possibility over actuality, Heidegger enables the creation of a living archive of what has not been, nor can ever be, accomplished in keeping with the domineering mission of thinking, an archive of another world, which is not put as a harness on the tamed earth.

The essentially belated disclosure of possibilities at the end of "today's" day is patently Hegelian. What is not at all dialectical is the mechanism that makes this disclosure happen: instead of relying on the retrospective standpoint of a mature concept, Heidegger urges thinking to unclench its grasp, reverting from the capture to the release of the world and of the earth alike. If there is still any hope left, it has to do with the world letting itself go and freeing the earth in the same gesture. Only in letting go of itself does the world remain possible.

Heidegger's thinking release will not save us. But, without it, we are more lost, more devastated and devastating than we are with it.

3

PLUS DE RESTES: REMEMBERING JACQUES DERRIDA

The following text was occasioned by the tenth anniversary of Jacques Derrida's death, marked on October 9, 2014.

With remarkable insistence and uniformity, recent commemorations of Jacques Derrida, occasioned by the tenth anniversary of his death, have shown preoccupation with the question of remains. Exemplary in this regard is the article "Que reste-t-il de Jacques Derrida?" (What remains of Jacques Derrida?) by Michael Behrent and Héloïse Lhérété in *Sciences Humaines*.[1] A great deal of anxiety underlies queries such as the one Behrent and Lhérété are making. It doesn't even matter if the queries are full of reverence or contempt: while Derrida-philes worry that what remains is too little and that the intellectual influence the great master of deconstruction used to exert is fading, his detractors bemoan what they consider to be the undiminishing clout of his thought across the humanities and, to a lesser extent, the social sciences and legal studies.

A panoramic look at "what remains" is part and parcel of the work of mourning that takes stock of and accounts for the traces left by the lost object. Derrida himself commenced the work of mourning his own absence in his written texts and carried this work through whenever he invoked, for instance, the issues of legacy, his testamentary desire, and the aspiration to leave marks on the French language. It is worth asking, then, who raises such concerns and how. Who has the right to mourn (above all, publicly) someone who, across virtually all his works, has been in mourning for himself, without end? Those who knew him personally? His disciples? Official legatees? No one?

An indication that something is going awry in the current propagation of accounts of and accountings for the mourned object is that "What remains of Derrida?" really means "What remains identifiable of his texts, his themes, or his influences?" The same motivating force unites tributes and diatribes that share the goal of categorizing, fixing, and passing a judgment upon the remains in question; in short, the goal of determining a legacy. As if ten years, or any number of years, for that matter, were sufficient to endow the survivors—all of us—with the privilege to do so and, on this basis, to define the future reception of Derrida.

If you are au courant with Benoît Peeters's recently published monumental biography, you will remember that the note Derrida had asked his son to read over his grave amounted to an interdiction of mourning: "He begs you not to be sad. [...] Always prefer life and never stop affirming survival."[2] This posthumous request does not go so far as Jean-François Lyotard's explicit prohibition (unless it is a constative or a normative, rather than a prescriptive statement), "there shall be no mourning [*il n'y aura pas de deuil*]," which Derrida reported in his funeral oration for the dead friend. "This was about ten years ago," he adds[3]— the same time span that separates us from his own death. "There shall be no mourning" does not spell out either utter oblivion or persistent melancholia. Instead, it signals, by a sort of *via negativa*, a mode of relating to a body of thought, of work, and perhaps to a life that does not rely on any mechanisms for identifying the remains in question.

In an entirely different context (in his defense of another friend, Paul de Man), Derrida gives a name to the unidentifiable afterlife of thought, work, and life "itself." That name is biodegradability. "The worst but also the best that one could wish for a piece of writing," he writes, "is that it be biodegradable. And thus that it not be so. As biodegradable, it is on the side of life, assimilated, thanks to bacteria, by a culture that it nourishes, enriches, irrigates, even fecundates but on the condition that it lose its identity, its figure, or its singular signature, its proper name."[4]

Isn't this what one should wish for deconstruction and for the proper name "Jacques Derrida," too? Isn't this what is already happening (and has been happening for quite some time before 2004) to deconstruction, despite all the efforts at circumscribing its remains? Wouldn't its biodegradability imply the deconstruction of deconstruction, not as a formal meta-critique turning the tables on deconstructive hypercriticism, but as the intensification of the work or the play Derrida has favored in texts that bear his singular (and also iterable) signature?

In this scenario, both more and less remains than we think: *plus de restes*. Bereft of discernible boundaries, a piece of thought or of writing is everywhere in its effects and nowhere to be found as such. Thanks to a strange tradeoff that does not fit within the limits of economic logic (the best and the worst coincide), the singular signature and the idiom step aside, give themselves up, diffuse into the soil of culture so as to make future thinking possible. To me, the acceptance of such a condition betokens the fecundity and, especially, the maturity of thought, freed from schools, circles, and societies that invariably mutilate it by forcing it to conform to a recognizable and simplified image: an idol. It could well be that biodegradability is how the gift of thinking is given in Derrida's aporetic sense of the gift, according to which neither the giver nor the receiver is aware of the giving.

So, why does biodegradability stand both for the best and for the worst one could wish for a piece of writing? Derrida hints that a non-biodegradable remainder offers resistance to the assimilation of a work into the cultural milieu and that it constitutes "the remains that remain to be thought," as he notes on the same page of the essay. It is this remainder that is lost when thought or writing dissolves into compost for ideas. But what if something other than the remains of singularity and idiomaticity remain to be thought? What if these remains belong to the anonymous stratum of existence, which has always been inaccessible to metaphysics?

By "anonymous existence" I do not mean Heidegger's *es gibt* or Levinas's *il y a*—the thought that "it gives" or "there is" being. For me, the names, the synonyms, of this anonymity are plants, growth, and *phusis*. After much energy has gone into deconstructing the contrived opposition between "nature" and "culture," now is the time to return to *phusis* after deconstruction and to affirm its burgeoning emergence that offers an unacknowledged model for the thinking of being as a tireless giving of itself. What is necessary, by implication, is a return to plants, the growing beings par excellence, the metonymies of this emergence. I doubt that Derrida would have subscribed to (literally: put his signature or his name under) this turn or return; however, faithfulness to a legacy does not require that one keep reiterating the same doctrinal contents or that one repeat the theoretico-practical gesture that led to these contents' enunciation. If assessments along the lines of "What remains of Derrida?" are to be worthwhile, they ought to pay attention to the oft-times unexpected venues for thinking and being opened by deconstruction as it survives, unrecognized, its still continuing "descent" from the biodegradable to the biodegraded.

4

THE PHILOSOPHER'S BEARD

This 2015 essay deals with a markedly gendered physical attribute of (some) philosophers.

Whenever I attend conferences and symposia, male philosophers, who unfortunately still represent a vast majority in the profession, are very easy to spot. Usually, they consume exorbitant amounts of caffeine and sport beards of various shapes and sizes. Even colleagues I recall having a clean-shaven look fairly recently, give free reign to this particular secondary sexual characteristic. It is in such moments that I wonder if philosophers ever consider their beards philosophically.

One obvious objection to special attention paid to the philosophers' beards might be that it is currently fashionable among men in the general population to sport facial hair. But in my purely subjective experience, the trend is greatly exaggerated among male philosophers, who, as philosophers, should be expected to think for themselves, and so be highly resistant to the current fashionable trends. Do they resort to the beard out of some repressed anxiety, using it as a physical marker of sexual difference at a time when more women enter the discipline? Do they simply perpetuate a stereotypical view of the philosopher's countenance, going all the way back to Socrates and Plato and now engrained in popular culture? Or is something else going on here?

Take, for instance, Hegel. For the (clean-shaven) German thinker, human beings are physically distinct from animals, such that our flesh is almost entirely spiritualized. Instead of thick scales and fur covering our bodies, our skin is largely exposed to sight and touch, supplying further evidence for the increased sensitivity of our bodies. On this view, fur and hair are remnants of plant nature, archived in us and, to a greater extent, in animal corporeality. They are reminiscent of vegetal growth, just as, from Antiquity to (bearded)

Michelangelo's Renaissance, the forest symbolized the hairy cover of the living organism that was our planet.

If we treat this half-forgotten analogy seriously, then the act of shaving becomes imbued with strong philosophical connotations. "Clearing" vegetation—an age-old practice that has led to the disastrous rates of deforestation worldwide—has been conventionally represented as a struggle for civilization and the Enlightenment, which is also a fight against the stubborn and dense growth, obscuring the surface of the planet. Not surprisingly, shaving has been historically linked to the same philosophical-political cause. Suffice it to mention Tzar Peter I of Russia, who, on the cusp of the eighteenth century, imposed a "beard tax" on the male part of the population in an attempt to modernize the country. His effort to integrate Russia with the rest of Europe and to spread the values of the Enlightenment across his domain was, to his mind, consistent with making popular the clean-shaven look of the Continent's "civilized" portion.

Curiously, the civilizing and humanizing endeavor, expressed in part in the elimination of beards, brings men a tad closer to women whose secondary sexual characteristics do not include the growth of facial hair. According to the unarticulated logic of the Enlightenment, women, with their skin more exposed and bearing fewer traces of animal and vegetal nature, will have been more fully human than men, who—should they choose to shave—face a daily struggle against the plant or the animal in (or on) them. As a result, Hegel's own association of women with passive plants, as opposed to the active core of men pictured as animals, becomes internally contradictory and collapses under the weight of physiological, hormonal evidence. Sexual difference is transformed into a differential in the degree of civilization that perverts the established hierarchy, or, better, subverts it from within.

I doubt that any of the above passes through the heads of my colleagues when they shelve their razors and other shaving implements. But, assuming that anything other than one's image and public persona were at play here, attachment to facial hair would have indicated physical resistance to the staple Enlightenment values of absolute visibility and transparency. It would have been a way to embrace the animal and even vegetal heritage, which, when left untamed, proliferates on the surface of our bodies. And, more significantly still, it would have signaled the return of the body in or to philosophy that has for too long been repressing corporeality and fetishizing an abstract, nearly disembodied mind.

At the level of sheer appearances, though, the beard continues to set male philosophers apart from their female colleagues. Symbolically linked to an exclusionary and chauvinistic tradition, this aspect of physical appearance may unconsciously signal the obduracy of old dividing lines between philosophers and non-philosophers. This, too, needs to be thought through at a time when women continue to face challenges on their path to gaining admission within an academic field traditionally dominated by men.

Be this as it may, the beard hides the mouth, the organ Hegel marveled at for its combination of the lowest and the highest functions; it is the first station in the digestive process and the last step in vociferation, the production of speech. Food goes into it, words come out, and, in men who do not shave, all these comings and goings happen under the cover of the beard. In the case of bearded philosophers, which one of these phenomena do they try to occlude? The intake of food? Or the output of the word, of *logos*, the main tool of philosophical trade? In reality, they cannot help but obstruct both: the most basic materiality of organismic self-reproduction and the highest ideality, still rooted in corporeal existence, of the voice, of speech, of articulation, of the word.

What we have before us here is strong evidence for the ambivalence of the philosopher's beard at the same time negating and reaffirming the tradition. Entangled in the beard is the embrace of human "vegetality" and the conventional image of a male philosopher, a reaction against the Enlightenment values of clarity and transparency and an attempt to follow present-day (or millennia-old) fashions. At the apex of physicality, this ambivalence is expressed in the simultaneous concealment of speaking and eating, of philosophizing itself and of nourishing the philosopher's body. Yet, when it comes to food and drink (as well as, say, tobacco), the beard often bears the traces of the substances ingested by its owner. Perhaps, then, sporting it veers on the side of materiality, marking the philosopher's face as human, all too human. That is to say, as animal and even vegetal.

5

NATURALIZE THIS! ANALYTIC PHILOSOPHY AND THE LOGIC OF REACTIVE NEUTRALIZATION

> Contemporary philosophy is scarred by the split between the "analytic" and "Continental" traditions. The following essay, from 2010, reflects on the disciplinary crisis triggered by this split and on certain disingenuous features of philosophy's "analytic" segment.

Over the past twenty years, a sizeable segment of analytic philosophy has been openly promoting naturalization, a process that has implicitly defined the goals of this philosophical strand since its very inception. The object of naturalization is so diffuse as to include epistemology and phenomenology, jurisprudence and education, power and responsibility, and, in effect, any human phenomenon whatsoever. The extent of this devastating trend makes it a good candidate for close critical scrutiny, which can help us diagnose the condition of analytic thought, structurally incapable of a sober self-assessment, and to explain its pernicious political consequences.

What distinguishes naturalization in its multi-faceted manifestations is, above all else, its reactionary core. Unlike naïve or pre-critical naturalism, it defensively responds to the denaturing of ways of thinking, socio-political processes, and economic institutions in nineteenth- and twentieth-century philosophy, sociology, anthropology, and literary studies, among other disciplines. It is not difficult to pinpoint the reasons behind this quintessentially conservative reaction: critical thinking is threatening to the status quo, with which much of analytic philosophy is aligned. At the institutional level, a glance at the predominance of analytic philosophers in the American academic universe is sufficient for one to realize just how useful they are in their function of providing ideological justifications for the perpetuation of political and economic injustices both inside and outside the university. Those wishing to maintain this skewed balance of power understand that a simple dismissal of threatening currents of thought is not an effective strategy; rather, they strive to appropriate, domesticate, and finally neutralize everything and everyone deviating from their self-proclaimed norm. Most remarkably, however, the reactionary tendency toward naturalization proceeds in the name of overcoming the split between opposing styles and approaches to philosophy, reconciled on the grounds of an already hegemonic school of thought.

The type of reaction that naturalization exemplifies falls under the classical Freudian category of disavowal, the simultaneous acknowledgement and repudiation of a threatening piece of reality (such as, in the case of psychic reality, sexual difference). After a drawn-out process, a norm is established as natural in a development that, at once, recognizes and neutralizes the previous deviation from the norm. Although it has little to do with nature, the newly recovered normativity vehemently refuses its association with any social, legal, or political conventions, thereby effacing the evidence of its own, rather sloppy fabrication. On this pretext, naturalization attempts to bring critical thought back into the fold of positivity, invariably modeled on the positivism of the natural sciences, and, in so doing, to assimilate everything in its path to the thinking of identity and instrumental rationality.

Under the title of naturalization, which, as already mentioned, remains suspiciously vague in a discourse committed to the rigors of argumentation and clarity of expression, we encounter nothing more than a reductive comprehension of nature as a set of empirically verifiable causal relations and quantities of force. The proponents of this approach are loath to ask the crucial question, "What is nature?" which is bound to influence their operations. They have no intellectual resources at hand for raising the question regarding the meaning and the being of nature: to "analyze" the latter, to break it down into quantities and relations of effective causality, is tantamount to giving a one-dimensional response to a question that has not been raised. But nature is not an analytic concept and, therefore, it fails to shore up the machinations of naturalization in a meaningful way.

Abdicating the traditional prerogative of philosophy, the analytic worshippers of reductive naturalism inherit the modern scientific conception of the natural, which serves as the gold standard for reducing phenomena from all domains of human activity as well to hard data. This is not to say that the category of nature as such holds no promise for the thinking of ontology rid of its metaphysical overtones; indeed, much work is still to be done in this field. Only when "nature" is thus denatured (that is, released from the scientific-analytical straightjacket) will it be desirable to bring it to bear upon the various domains of human life. Naturalization will then play a role at odds with the one it fulfills today: it will signify the liberation of the political, cultural, cognitive, social, and economic spheres from the double yoke of positivism and anthropocentrism.

Without such an alternative, reactive naturalization will continue massively transcribing the oppressive categories of the past, shorn of their content, into the tyrannical ontology of the present. A change of paradigm in modern science saw the decline of the idea of nature, which had remained largely unchanged since Greek Antiquity, and the rise of the mathematico-algorithmic model of reality. Considering that the premodern teleology of nature used to justify the static and oppressive socio-political arrangement of its time, a dire need for re-grounding domination in the new scientific order becomes palpable. All this will come as no surprise even to a novice reader of the critical theory of Frankfurt School. But it is still worth examining the details of the discursive deployment of nature with the view to outlining the general strategy its practitioners adopt.

In the legal domain, "natural law" presupposes a premodern, teleologically inflected ontology of nature. The universal principles inherent in this image of the law are metaphysically determined in keeping with the objectively fixed hierarchy of ends and, later on, the standard of truth emanating from the word of God. Legal positivism is usually taken to be the exact opposite of natural law, while it is, in fact, but the end result of dissolving the old notion of "nature" into the categories of modern science. The naturalization of jurisprudence, with the attendant conceptualization of legality on the basis of cause-effect relations, enables the same legal oppression as the one that marked the premodern notion of law. What both approaches have in common, then, is their insistence on the objectively fixed (or, at least, fixable) meaning of law, defined by the teleology of nature, by the word of God, or by the impoverished notion of effective causality. Each of these precludes the legal hermeneutics that potentially unhinges the meaning of law itself, opens up a multiplicity of interpretations and, as a result, de-naturalizes systems of social domination.

In the realm of thinking, naturalization aims to extinguish the critical impulse that emerged in Kantian and post-Kantian philosophy after an objectively constituted epistemology had disintegrated. The natural order of thought depended on formal logic, the thinking of identity that in its stability and immutability was said to resemble the eternal repose of the gods. The Hegelian critique of formal logic was only the beginning of the de-naturalization of thinking explored, in various ways, in twentieth-century philosophy by Theodor Adorno, Emmanuel Levinas, or Jacques Derrida, to name but a few emblematic authors.

As a reactionary response to these philosophical inroads, the recent rush to "naturalize the mind" takes two distinct forms. First, it can revert to the premodern paradigm and insist on the absolute tyranny of formal (and symbolic) logic. Second, it can bow to the modern conception of the natural and translate mental processes into the scientific terms of cognitive science.

The second option is the one that the school of "naturalizing phenomenology" wholeheartedly subscribes to. Here we have a case of naturalization that is extremely revealing with regard to the goals of analytic philosophy and to the bizarre ways in which it disfigures its naturalized objects. Husserlian phenomenology is impossible without a rigorous reduction, bracketing, or suspension of the natural attitude—whereby our world is pre-comprehended on the basis of unexamined common sense—and of scientific conclusions about reality. To naturalize phenomenology is to rob it of its critical bent,[1] in the absence of which the phenomenological method and the provisional conclusions it points toward turn insipid and get neutralized. Forcing phenomenology into the cast of cognitive science and mathematically modeling phenomenological descriptions, the movement of naturalization destroys Husserl's aspiration to reground human knowledge and the sciences on philosophically sound foundations. As a consequence, the very crisis of the European sciences that Husserl tackled in one of his most important late works is intensified: the means for overcoming this crisis are subordinated to abstract models and the exigencies of positivism, marking modern scientific rationality.

A glimmer of hope in any crisis is not that, once it passes, everything will return to "business as usual," but that it will radically transform the system it affects. In other words, a crisis potentially de-naturalizes ossified beliefs and practices by showing that they are neither universally applicable nor indispensable. The crisis of the global markets puts in question the assumption that capitalism is the natural mode of organizing economic life; the crisis of liberal democracy shatters the dogmatic view that it is the natural way of coordinating political life; the crisis of philosophy, split between the "Continental" and the "analytic" strands, defies the idea that there is a single method or path leading us to an objective truth, or that there is an objective truth to begin with.

At the same time, crisis is not a panacea from uncritical modes of thinking and action, since it can undergo a meta-naturalization (and, hence, be neutralized) as soon as its threatening, destabilizing effects are used to consolidate the same logic that brought it about. Instead of occasioning a sustained reconsideration of the bases of economic life, the most recent crisis of capitalism has spawned the most unjust excesses of this economic system to date: wealth is further redistributed from the bottom up, from ordinary taxpayers to the banks and transnational corporations. The crisis of philosophy has similarly led to the entrenchment of its most uncritical elements, claiming the right to represent this entire academic discipline today, the elements that dare naturalize and neutralize critical thought itself. In the face of this onslaught, our response has to be both harsh and rigorous; we must lucidly spot the enemies of thinking—those opposed to its free flourishing—and set them in the context of intellectual history, wherein they will finally appear as the reactionary dogmatists and civil servants of the sociopolitical status quo that they are.

6

JOKES AND THEIR RELATION TO CRISIS

This 2011 text contemplates the role of humor in crisis situations.

In June 2011, Republican presidential candidate Mitt Romney found himself under attack for a joke he tried to make at a meeting with a group of unemployed people in Tampa, Florida. "I am also unemployed," Romney announced, insinuating that the job he lacked was the presidency.

Romney's mistake was to have ignored the meaning of the economic crisis, including the class-based divisions and anxieties it had aggravated. His statement of identity and identification ("I am *also* X") achieved the exact opposite of the desired effect, underscoring the unbridgeable gap between the "unemployed" multimillionaire and the out-of-work Floridians. But in a sense, the joke worked, though not in the way Romney intended: it showed, above all, his own cluelessness. The joke was on him.

The subsequent moralizing responses of Romney's critics were remarkably uniform. They boiled down to the admonishment that the crisis is not a laughing matter, that poking fun at unemployment is disrespectful to the unemployed, and so on. But what if, on the contrary, humor and crisis share a common provenance? What if humor invariably germinates in response to a crisis, as a reaction to the excessive splits between us and our social, political or economic reality; or to divisions within us; or to rifts within reality itself? If so, then laughter, instead of mending the multiple fissures of the crisis, only further accentuates them, makes them finally and fully what they already are. Humor is not, as some believe, a coping strategy or an outlet for the frustrations that cannot be expressed in any other way. Or it is not just that. At its best, it is the self-consciousness of crisis.

Without changing anything in "objective" reality, humor permits reality to laugh at itself, and, therefore, to stand outside of itself, to become other to itself.

At the core of the word "crisis" (which, like "critique," is derived from the Greek verb *krinein*, connoting separation and division) is this hyperbolic distancing from itself of the entity that undergoes it. Humor might well be the mechanism that triggers the transition from one sense of separation to the other: from crisis to critique.

Consider the following joke: "In America, banks rob people because that is where the money is!" In the reversal of commonsense, which is but another name for ideology, this pithy remark echoes the rhetorical question of Bertolt Brecht, "What is robbing a bank compared to founding one?" Implying that the financial system as a whole is the institutionalized apparatus of robbery, the joke instantaneously pierces through the veil of ideology. Now, a humorous accusation often stops at just that. It accepts the target of its critique as a brute fact of life. While no one supports the conduct of banks stealing from the people, laughter confirms that the audience fatalistically accepts this political-economic reality. *C'est la vie!* is the secret motto of humor.

Still, fatalism is only one piece of the puzzle here. Also at stake in the logic of crisis is nothing less than human finitude: our limited capacity to manipulate the future, aging, and our impending death. Suffice it to look at two popular jokes to realize that the issue of time shadows everything that has to do with the crisis:

1. What's the definition of optimism? An investment banker ironing five shirts on a Sunday evening.
2. A man goes to his bank manager and says, "I'd like to start a small business. How do I go about it?" "Simple," says the bank manager. "Buy a big one and wait."

In each case, finite time—the time that is about to reach its term; the future, more threatening than ever before—is the object of concern, animating the joke. If crisis puts us face to face with the unknown, it cannot help but provoke extreme anxiety about the uncontrollable future.

How does humor relate to this future and how does it cope with this sentiment? Far from assuaging the anxieties stimulated by the unknown, it is a symbolic device that enables human beings collectively to confront our own ageing, death, and the limits of our sexual, social, political, and economic realities. In the insecure employment (even) of the investment banker and the rapid melting away of savings, we recognize ourselves in the present and, more importantly, in the future, in which we might be capable of forming a basic bond of solidarity.

Thus: (1) the temporal fissure between the present and the future is the site of the crisis; and (2) humor puts this divide under a symbolic spotlight. But who, exactly, laughs at whom when the temporal structure of the crisis is made visible? Is it the present that laughs at itself? Does it chuckle at its grim future? Or is it our future, laughing at us in the present?

Laughing at ourselves, at the various crises in which we are steeped, is laughing at our irremediable weakness, the feeling of being overwhelmed and crushed by the future. This would suggest a straightforward interpretation, whereby the present laughs at the future, or, at any rate, at its own fear of what is to come. To make fun of the future is to put it under our control, if only for a brief instant of a shared explosive laughter, by conquering the fear of the unknown.

But humor's mastery is insecure; it reveals the strength *of* weakness, which is our capacity to face our own predicament without dissimulations, false reassurances or unrealistic expectations. Instead of endeavoring to master the future, we plunge into the chasm between what is to come and the present, deepening the crisis in ironic self-consciousness.

Jokes, even if they are meant to be shared against the background of common cultural and linguistic presuppositions, are the sites of friction: herein lies their affinity to the situations of crisis. Internally contradictory, they spell out the fatalistic acceptance of reality *and* discontent with what has been thus accepted; powerlessness *and* a new empowerment; a confrontation with the terrifying future *and* an openness to the weakness it portends. Humor is one of the best outlets for expressing the crisis, itself a result of internal contradictions and divisions that have become unsustainable in their current state.

Paraphrasing Heidegger, we might say that the essence of humor is nothing humorous; it is, rather, the separation, variously called "time," "self-consciousness," "critique," or "crisis," of the I from itself and from the world I live in. When humor responds to a crisis, it reverts to its own essence, launching a tacit critique that retraces the divisions and contradictions, out of which the crisis has erupted. But, while the essence of humor is nothing humorous, this should not hold us back from having a good, hearty laugh.

7

POSITION AS A POLITICAL CATEGORY: PHENOMENOLOGY AND THE EROTICISM OF POWER

This essay was written as a companion piece to my book, Political Categories: Thinking beyond Concepts *in 2019.*

Two motivations guided my writing of *Political Categories*.[1] The first, revealed in the preface, was the need to outline the hallmarks of politics in the face of a neoliberal assault that purposefully and systematically conflated this domain of human activity with economics. The second emerged in the measure, in which the book's argument unfolded: the desire to recover political experience, or, more precisely, politics as a matter to be experienced in its institutional and revolutionary, formal and informal, variations. Succinctly put, I aimed to elaborate a phenomenology of *res publica*, to which all things political, and even those usually set outside the political sphere proper, referred.

While the theoretical pillars for my work were Aristotle's and Kant's reflections on the categories, my constant interlocutor—a philosophical *frenemy* of sorts—was Carl Schmitt. It may not be obvious, but the very title of the book is polemical: it suggests that political categories offer a better, more nuanced, vigorous, and flexible approach than Schmitt's "concept of the political." I do not, in fact, believe that there is such a thing as *the political*. To my taste, the term is too vague and abstract to be meaningful; it perfectly matches an eviscerated politics, shorn of *res publica* and its corresponding experiences. Together with the concept, the political dispenses with the irreducible plurality of politics, the plurality we may access by resorting to the categories that do not, taken singly, exhaust the complex meaning of the things they comprehend. I much prefer *politics*—a noun saddled with multiple and frequently contradictory senses that are still connected, in one way or another, to actual practices and everyday life.

In the brief remarks that follow, I would like to zero in on one of the Aristotelian categories and, in the spirit of the phenomenology of *res publica*, elaborate its relation to the question of sexuality and the eroticism of power. That category is positionality. The germs of the elaboration are scattered throughout *Political Categories*, especially in the chapter dealing with the state. Here, I want to flesh it out further in light of the sexual ontology of individual bodies and of the body politic.

When we hear the word *position* in a political context, we immediately interpret it in terms of a demand to situate oneself on a spectrum of practical options in response to a problem or question. These options are not the expressions of ideas but of worldviews and opinions that, typically polarized, become so banal (if emotionally charged) as to turn into something like caricatures of themselves. What is your position on abortion? On universal healthcare? On migrants? The clashing camps that endorse diametrically opposed positions are utterly predictable in their argumentation, and much of the 24-hour news cycle is based on replaying them over and over again.

A more phenomenologically grounded approach to political positionality has to do with the division between the right and the left. Although these designations refer to the ideological differences among the adherents of various positions, they stem from the spatial arrangement of the French National Assembly in the aftermath of the 1789 Revolution. There, the deputies sat to the two sides of the president's chair, to its right or to the left, according to their allegiance to the king or the revolution. A lived orientation underpinned the division: positions of bodies in (a politicized) space were imbricated with political positions and reflected in miniature a complex position of the body politic at the time.

In the absence of a monarch (and, later on, with the decline of centralized authority), positional markers outlive their initial phenomenological *raison d'être*. To us, the point of reference in the question *to the left and to the right of what?* is the center, a fluctuating agglomeration of the widest possible consensus situated between ideological extremes. Consequently, the recent collapse of centrist parties and coalitions all over the world renders the political positions "left" and "right" meaningless. It completes the process of decoupling them from their experiential foundations—the process, which has been ongoing for over two hundred years that have elapsed since the French Revolution.

What interests me, nonetheless, is the embodied positionality of political groups and their members. The state, for one, is, at the semantic and symbolic levels, a political unit in a standing position. Its status is that of something erect. Confronted with the state, its own citizens must assume a passive position: they must lower themselves in comparison to the power transferred to and accumulated in the institution. The citizens are on the receiving end of the state towering over them, unless the sovereign fails to fulfill its protective function, in which case rebellion is justified. That, in a nutshell, is Hobbes's take on the matter.

Evidently, the position of the state is phallic. But its never-ending erection is as much an instrument of authority as a seed of its demise. Always standing, the state does not respect sexual and cosmic energy cycles: the period of relaxation that follows the phase of excitation; the rotation of planets, colloquially and experientially expressed, for example, in the rising and setting sun. The state cannot afford to change positions, despite all the rhetoric surrounding the decentralization of its power. It is a faithful inheritor of the phallic-solar fetish in politics—of the Sun King who morphs into the sun that "never sets over the British Empire." Never setting and never sitting, never going down, state sovereignty is all-powerful and, by the same token, powerless: exposed in its perpetual genital display, even if this exposure is meant to be blinding (and so, concealing a great deal) in its solar brilliance. Compared to sitting or lying down, standing is also a position that is less stable, less supported by a solid stratum, which, in the last instance, has to do with the earth. As such, the one standing is more prone to overthrows, confirming a piece of common wisdom that proclaims, "What has come up must, at some point, go down."

While, in relation to the state, citizens assume a passive and lower position, as its representatives, they stand up and, regardless of who they are, are rendered phallic. We should interpret "a standing army" in this vein. The admission of women into army ranks is, therefore, not quite an advance in gender equality, but another mode of submission—by identification with phallic authority. Curiously, Trump's objections to transgender people serving in the US military (overturned at the very start of Biden's term in office) impose a discriminatory, if, maybe, also an enabling limit on this process of identification. Trump denies a particular group a "standing" (the legal translation of the state's erection); he puts in doubt the very possibility that it can stand, that it can threateningly expose itself—as who or as what? Effective resistance, then, should not buy into the typical liberal demand to accept previously excluded others into the expanding domain of a proudly erect center. Instead, it should take another position (sitting, or lying down: we will turn to these in a moment) from which to undermine the dominant standing.

In Emmanuel Levinas's occasional forays into political thought, the question of position is quietly operative behind the scenes. His view of political life makes solidarity prominent in a collective shoulder-to-shoulder stance of the citizens.

To extrapolate from this image, although members of a polity do not expose themselves to each other, they are still standing and standing up. Before whom?—that would be Schmitt's question to Levinas. Does being side-by-side make sense without a confrontation (perhaps, only a potential one), a united front the group presents against another group? For Levinas, it is, curiously, the ethical relation, rather than politics, that hinges on a face-to-face confrontation, asymmetrical as it may be. In facing and being faced with the other, the standing of the parties in this strange relation is unequal, if not altogether incommensurate. It is a standing, nonetheless, characterized not only in vertical terms, with reference to height, ascendence and transcendence, but also with regard to the aiming (*viser*) of the face (*visage*). Levinas himself concurs that any part of the body can be a face—say, the hunched back of a grieving mother standing in a line of people eager to receive news of their loved ones arrested by secret police in the Stalinist purges.

Sitting seems more pacific than standing, and more stable at that. The seated bodily position is also ambiguous compared to standing, because, unlike the latter, it does not reveal the sexual organs of the one who sits. To us, hypermoderns, being sedentary sounds like the negation of mobility, a symbol or a symptom of the feudal past, of a fixed politico-economic order where the lords and their serfs were attached to particular plots of land. Human settlements are fundamentally agrarian, inseparable from the places of plant cultivation and phenomenologically aligned with the sessile nature of vegetal existence. That said, sedentariness and sessility are the positional sites of intensive movement, of movement in a place, with growth, decay, and metamorphosis for its salient modalities. In other respects, too, sitting is far from passive. The resistant position par excellence, it negates the unfettered mobility of capital and the more conditional mobility of labor glorified in our era. And that is not even to mention sit-ins as effective forms of protest, or "squats" that occupy disused buildings and repurpose them for housing.

Physically more stable and sexually indeterminate, the sitting position challenges the verticality of the human body and its potency, doubly coded in terms of political power and masculine sexuality. The upper portion of the one who sits retains a vertical stance, while the lower part is horizontal in relation to the floor or the ground. Sitting is a physico-symbolic fold imbued with mystery and vegetal eroticism, a crease of activity and passivity, oppositionality and receptivity.

The crease of sitting is undone in a lying position that unreservedly embraces the kind of horizontality capable of subverting power hierarchies. Lying down is not just a position of the corpse (which is, actually, the most difficult and deeply energetic yoga pose, *śavāsana*), of what, with a great deal of irony, Kant called "the peace of the cemeteries"—even if it has been adopted in "die-ins" in the course of recent Extinction Rebellion protests. Nor is it the outcome of absolute submission. A horizontal position is that of grassroots movements, as well as anarchic networks, that, at their core, oppose the state's stance, its standing as the erect embodiment of hierarchical power.

The paradox of horizontality is that it must oppose verticality without opposing it, without engaging in a standoff, which would draw it into the orbit of what it resists. From the standpoint of someone or something standing, it would appear purely passive and available, up for grabs, ready for the taking. But the lying position renders the one who assumes it both exposed (from a different side than that of the genital exposure embodied in the ever-erect state) and elusive, unwilling to engage with the powerful stance on its terms. To lie down is to receive the most support from the ground, which, in the end, cannot be appropriated, despite being carved up into plots of land and territories forming the basis of possession. It is to be suffused with this non-appropriability and to impregnate the ideality of power with the gravity of materiality.

A revolution, as I note in *Political Categories*, is the inversion of a position occupied by the entire body politic. In this respect, revolutions are more radical than revolts that signal slight shifts in the body politic, but they are less radical than resolute insistence on horizontality. Revolutionary overturning may put the collective subject legs up, or, on the assumption that the standing of the status quo is itself perverse, invert it the other way, restoring the morally and physically upright stance. "Taking power," revolutionaries re-eroticize it still with the view to the phallic fetish. That is, arguably, why historically accomplished revolutions invariably fail. Their historico-phenomenological contexts beg a broader question: are all varieties of power that attain hegemony necessarily circumscribed to the vertical axis of spatio-sexual experience?

8

THE POWERLESSNESS OF PHILOSOPHY

Written in 2020, this essay contemplates the role of philosophy with respect to existing power structures and ordinary people's lives.

A towering figure in the study of ancient thought, Pierre Hadot outlined three aspects of pursuing philosophy in Antiquity. Crucial to this miniature classification was the *context* of doing philosophy, rather than its methods, form, or content. Hadot then correlated the three aspects to archetypal figures, namely "the philosopher living within his school, the philosopher living in the City, the philosopher living with himself (and with what transcends him)."[1]

To non-specialists in ancient thought, the distinction may seem odd, to say the least. Don't most philosophers live in cities? Don't they all belong to one school of thought or another? And how are we to envision "the philosopher living with himself" or with herself, if not through a broken prism of intellectual biographies and autobiographies?

Let's take a closer look at the first and the third of Hadot's points. Living in a school of philosophy is a proto-monastic experience, replete with a venerated spiritual master and a strict organizational structure that is set up at arm's length from the world. A philosopher living with herself or himself engages in what Hadot calls "spiritual exercises," including examinations of conscience, meditations, and "the inner and mystical life."[2] Both of these aspects of doing philosophy seem hopelessly outdated, surviving in a hardly recognizable form, at best. The proverbial ivory tower of the university is a distant and perverted echo of ancient schools, while spiritual exercises in the commercial wrapping of yoga and meditation classes are hobbies only incidentally relevant to obtaining tenure or dealing with the stress of the publication assembly line.

More curious is the fate of "living in the City." With respect to this contextual facet of philosophizing, Hadot writes:

> The philosophical schools never renounced exercising an active influence on their fellow citizens. The means mobilized to achieve this end are certainly different. Some philosophers attempted to effect direct political action, dreaming of seizing power. Others contented themselves with counselling the rulers. Others still put themselves in service of the city, in giving lessons to the ephebes or in attempting to rescue it by becoming ambassadors. Others hoped to make their fellow citizens comprehend what the true life is, proposing to them the example of their own life. All, in fact, think to change the way of life of their fellow citizens.[3]

Hadot confirms that philosophy in Antiquity is quintessentially ethical and political. The effects philosophers intended to have on their fellow citizens and on rulers are not belated additions to their theoretical programs. These hands-on consequences are at the core of philosophizing, whether the philosopher lives in the city or in a quasi-monastic setting. And it is here that distinctions start getting a little blurry: no matter how secluded a philosophical school, it aims to intervene into public life, just as no matter how singularly individual a spiritual exercise, its ultimate goal is to lead others by example.

Taken together, the three dimensions of philosophy imply that this type of human pursuit is simultaneously personal and public, esoteric and exoteric, in light of the groundwork its desired ethical and political effects provide. This deep integration of philosophical theories and practices all but dissipates in our times.

Philosophers who do not have "ethical and political thought" as their specialization are largely aloof to the broader effects of their writings or thought experiments. The few, who still endeavor to counsel rulers, function as apologists for the status quo (a case in point is Bernard-Henri Lévy in France, though many others can be cited as well). Those teaching at Columbia, Harvard, Yale, or Stanford—and, therefore, de facto bearing responsibility for the preparation of future political cadres in the United States and other countries around the world—rarely assume the onus of this implicit responsibility. A vast majority of public intellectuals often reduce their messages to a set of therapeutic platitudes.

If, as Hadot contends about ancient philosophers, all of them "think to change the way of life of their fellow citizens," then the question for us in the twenty-first century is: why think at all in a world where any such change is *a priori* ruled out? Or do we only think that we are still thinking after a vital aspect of philosophizing has fizzled out?

Recall that the ethical and political power of philosophy in Antiquity is, far from an extraneous addition, part and parcel of the task of philosophizing. (Were it not, philosophy with its advocacy of change would have been nothing but ideology.) It is along these lines that Aristotle saw in political science the queen of all sciences. For the Greek thinker, the most important question was the question of ends: *what for? what good is it?* The question of ends was the question bearing on the good, even and especially in its most pragmatic and practical versions. The science called "political" was concerned with the common good, which, in its universal embrace of all other ends, was the most encompassing. That's why politics, for Aristotle, was the highest of sciences, situated above the study of nature (physics) or of household management (economics).

To follow the thread of the above example (which is, of course, more than an example), contemporary philosophy and everyday discourse assume the irrelevance of ends and berate Aristotelian final causes. The consensus emerging through a pervasive critique of teleology is that there are no fixed ends, with the attendant conclusion that completion and accomplishment boil down to shameful illusions. Nevertheless, one and the same end, tied to ever-growing profit margins, animates much of cultural, intellectual, political, and, of course, economic life behind the scenes of the decline of teleologies.

What we have before us is a blatant, albeit unacknowledged, contradiction: *there are no fixed ends, but everything hinges on and is made to serve the end of profit-making.* A similar contradiction looms large in twenty-first-century philosophy. Metaphysics, with its perennial claim that the world is reducible to a single idea, substance, or being, which, at the same time, transcends it, has been declared dead. Still, under the guise of postmetaphysical neutrality, we find the metaphysics of a technological (more precisely, the technicist) mindset, armed with numeric abstractions and algorithms for managing social and political life, patterns of consumption, and the like. In other words, *metaphysics has come to a close, but everything is imbued with the metaphysics of technicity.*

We are a long way from the three aspects of philosophical practice, as Hadot summarized them. But, instead of nostalgically bemoaning the current situation, let's ask: what if the powerlessness of philosophy were a blessing in disguise? After all, when philosophers are perceived as direct, or even indirect, influences on fellow citizens and the authorities, they are often violently rejected and, worse, are subject to persecution and death. Just think of the trial of Socrates, in the course of which the Athenian jury found him guilty of "corrupting the young," as well as of Seneca's condemnation to suicide by Nero, Hypatia's mob lynching, Thomas More's beheading, Baruch Spinoza's excommunication from the Jewish community, Mahatma Gandhi's assassination by a Hindu nationalist ...

Bearing this depressing history in mind, the powerlessness of philosophy, a discipline that is no longer taken seriously, is a blessing in disguise, not because committed philosophers can finally save their own skin—or, at least, not only because of that. In line with the other two contradictions we have already discerned, it means that *philosophers and philosophy have no power, but everything depends on the power of philosophy*. The metaphysical mindset triumphed after philosophers were convinced that it was over and done with; philosophy itself is poised to gain strength when the public at large and those in the positions of authority are persuaded that it is utterly useless.

The powerlessness of philosophers and of philosophy itself is the power of thinking in the face of multiple crises, when tried-and-tested solutions fail. It is the power of what appears to be futile in the present and of what, all of a sudden, becomes invaluable the moment the established order no longer works, say, due to the earth-shattering effects of a pandemic or of climate change. A power that, shorn of obvious effectiveness and immediate applications, is still consistent with the much-maligned highest ends of Aristotelian thought.

In 1847, Karl Marx published *The Poverty of Philosophy*, punning on Pierre-Joseph Proudhon's 1846 *The Philosophy of Poverty*, to which Marx intended to respond. Despite the promise of the title, Marx has little to say about philosophy in this work, developing instead the foundations for a future critique of political economy. Implicitly, though, his conclusion is that philosophy is poor as a result of its adherence to pure abstraction, which puts it on the side of exchange-value, oblivious to qualitatively differentiated ends. While alluding to this intellectual prehistory, "the powerlessness of philosophy" is not meant to be either exceedingly critical or dismissive of the philosophical endeavor. Quite the contrary, my point is that we should assume such powerlessness without a smidgeon of defeatism and in keeping with the other contradictions of our age. Once we do, philosophy might become something that truly matters.

PART IV

TECHNOLOGICAL UPHEAVAL

1

CHERNOBYL AS AN EVENT

> An essay written to mark the 30th anniversary of the explosion that destroyed Reactor 4 of the Chernobyl Nuclear Plant on April 26, 1986 and threw a gigantic plume of radioactive smoke into the atmosphere.

There are those exceptional instances, in which a date or a place stands for a particular event that happened then and there. "9/11" and "Chernobyl" symbolize, in a recognizable shorthand, two traumatic occurrences, two catastrophes that have since become watersheds not only for the American and European consciousness but also for world history and politics. Why the tendency to represent these events in such shorthand? Do we lack the time or the space to elaborate upon them in more than a word and a couple of numbers? Or is something else, something more profound, masking itself behind this marked brevity?

To symbolize an occurrence by means of a date or a place is a way to speak about the unspeakable. In so referring to traumatic events, we scarcely indicate them, keep them encrypted, or let them barely peer from the crypt of the unrepresentable. Perhaps, then, resorting to a psychological defense strategy, we cordon them off, condense them in and circumscribe them to a numeric or geographical inscription in order to tuck them into a far corner of the mind and not deal with them in all their excruciating, intolerable details.

One thing is certain: there is always a price to pay for the tricks of our psychic economies. Supplanting the event, no matter how traumatic, the date and the place become exclusionary. Or, even, egoistic. "9/11" names one event and, with this, erases multiple others, not the least of them the one now known as "the other 9/11," the day in 1973 when General Augusto Pinochet carried out a bloody *coup d'état* in Chile. "Chernobyl" not only designates the explosion of Reactor 4 at a Ukrainian nuclear power plant on April 26, 1986, but also expunges from collective memory the repeated massacres of its Jewish residents during the Russian Civil War and by the Nazi occupying forces during the Second World War. The shorthand goes hand in hand with the pretentious supposition that nothing else of what has ever happened or will ever happen on that day or in that place is worth remembering. The event prohibits all other events, including those that have already come to pass. It is like a black hole that sucks into itself the past and the future, along with vast stretches of space and language.

I have noted how events designated by a date or a place mark a day in a calendar or a geographical locale with their exclusive associations. Chernobyl is, in this sense, the most eventful of all the terrible events; it brands not only our conception of a given place but that very place itself, making it unfit for human habitation, extending a sizable "zone of exclusion," from which human inhabitants are barred, around it. This place that lends its name to an event loses every feature of placeness, to the extent that it no longer admits anyone into its midst, unless you are willing to risk your life by staying there (as some older residents of Chernobyl, who have refused to be evacuated in the aftermath of the disaster or who have clandestinely returned to their abandoned homes, have chosen to do).

There is yet another price to be paid: when the conditions of possibility for any experience, among which Kant counts space and time, blend with whatever is conditioned by them, they flip into the impossible. Chernobyl is this conditioned conditioning, an *atopos* where nothing can happen, at least for humans who have been permanently expelled from that non-place. It is the event of no event, the event that singularly negates the future, crushing it under the weight of nuclear eternity. Chernobyl, or that which now goes by that name, does not fill out the schema of experience because the radiation, which continues to ooze out of it, is not accessible to our senses. In all its elusiveness, the event of Chernobyl overwhelms our finite sensibility and nullifies the experiential schema itself. The conditioned conditioning turns out to be totally unconditioned and deadly in its absolute excess.

Nothing can happen in Chernobyl anymore, and a lot is happening there, just not for us. In this post-apocalyptic laboratory, the specimens of *Homo sapiens* are few and far between, while "wild" flora and fauna have made their comeback. The choice between reality in-itself, including its animal and vegetal instantiations, and reality for-us proves to be a false alternative; the post-Chernobyl world—a place now acquiring temporal, chronological connotations—is a reality for the plants and animals that live and die and, in some cases, adapt to an environment laced with high levels of radioactive elements. Still, however vigorously we dehumanize the event, the shadow of the future of no-future hangs over all living beings, and over the earth, in Chernobyl's exclusion zone. If reports are true that the plants harmed by extreme radiation are not decaying as they should, then the current vegetal and animal growth in the "radioactive reserve park" is unsustainable and might be short-lived, seeing that the soil will not be enriched through the normal decomposition of organic matter. A place arrogates to itself the title of an event to the detriment of its placeness and of eventhood as such.

As we begin working through the "event" of "Chernobyl," it is not advisable to dissolve the former in the latter, for instance, at the level of everyday discourse: "Chernobyl happened." A series of tough considerations await us. What have our technologies done to a place exemplary of human-induced devastation, if, certainly, not alone in suffering this disastrous impact? What has that place, converted into a vessel for the event, done to us? Where does the "event" of "Chernobyl" leave the consciousness rendered impotent by its effects? How does it influence human and other-than-human bodies?

A lonely word, which is moreover a geographical signpost, cannot cope with the thinking of the event. Taken in isolation, it prompts us to act out, rather than work through, nuclear trauma. In place of a designation for place, we sorely need both something less and something more than a word, namely *logos* no longer confined to human discourse and capable of bearing witness to the unbearable.

2

NUCLEAR MOURNING

Brief reflections, marking the 33rd anniversary of the Chernobyl nuclear disaster in 2019.

Like catastrophic global climate change, the nuclear age expands the circle of the "lost" objects to be mourned from the human to microbial, vegetal, and animal ecosystems, clean water, soil, and atmosphere, all the way to ideas and beliefs in personal and national safety, self-sufficiency, and sovereignty. Nevertheless, nuclear materials and effects are unique: at the extreme, they annihilate time itself. Their interference with the heterogeneous temporalities of living beings, communities, and environmental forces is derivative with regard to their negation of time. For instance, depleted uranium has a half-life of 4.468 billion years, in the case of U-238, or 700 million years, in the case of U-235, which, to all intents and purposes, approaches an eternity when seen from a human perspective.

Since a nuclear object does not pass away, refusing to become the past, it weighs heavily on the present, which it violently tears out of the order of time. With this, the nuclear object is converted into a material embodiment of trauma, psychic as much as planetary, a statically traumatic present and presence that cannot be metabolized, cannot be "digested" so as to free up time for the future.

How does one mourn a loss that resists passing away, that repels the past and the future alike? Nuclear energy generation and the disasters associated with it, the liquid and solid residues of power plants' "successful operations," military testing of atomic weapons, and the consequences of their deployment confront us with an object of mourning that, owing to its mode of being and temporality, thwarts mourning's work.

It seems, therefore, that nuclear mourning is an oxymoron. Confirming this impression, we are stuck at an interminable stage of melancholia, perpetually reopening the wound and letting it swallow our psychic, social, and ecological energies. The theological and metaphysical dream of eternity, perversely realized in the nuclear age, turns out to be a horrible nightmare, which has been the dark, unconscious underside of that dream well before the twentieth and twenty-first centuries.

One should not give up in the face of the impossibility to mourn in the nuclear age, however. A lot rides on efforts geared toward a reinvention of mourning today. In a totality of the melancholic condition where the work of mourning is paralyzed, there is nothing but denial, which only deepens the traumatic wound. Melancholic denial can assume several forms, ranging from arguments that atomic power production is safe to stubborn insistence on secure borders for one's personal and national home that overlook the uncontainable consequences of fallouts, among other regional and even planetary disasters. Political inaction at best and entrenchment in the most harmful practices at worst are the symptoms of this denial.

Nuclear mourning begins, for its part, from the depths of despair, but it should not end there. It strives to reinvent mourning by shifting its object from the dead or destroyed human or non-human other to the death of death. Rotting and decomposition have always been the aspects of mortality nourishing future lives and the future of life itself. It is these processes that are interrupted by nuclear materials, which resist decay, just as, on a different timescale, plastics, certain heavy metals, and so on clog the world and disturb planetary metabolism.

What we should mourn, then, is the death of death, which is also a vigil for mourning, a kind of meta-mourning, a mourning of and for mourning. For, only having worked through *its* loss, having mourned mourning itself, will we stand a chance in the difficult task of existence and coexistence in the nuclear age.

3

THE MEANING OF "CLEAN ENERGY"

Reflections on the senses of energy, written on the occasion of the signing of the Paris Agreement in 2015.

As the global conference on climate change is taking place in Paris, it is time to contemplate the meaning of "clean energy." In the West, the word energy is imbued with the force of deadly negativity. It is assumed, for instance, that energy must be extracted, with the greatest degree of violence, by destroying whatever or whomever temporarily contains it. More often than not, it is procured by burning its "source"— in the first instance, plants and parts of plants, whether they have been chopped down yesterday or have been dead for millions of years, the timescale sufficient for them to be transformed into coal or oil.

Without giving it much thought, one supposes that the only way to obtain energy for external heating or for giving the body enough of that other heat (namely, "caloric intake") necessary for life is by destroying the integrity of something or someone else. Life itself becomes the privilege of the survivors, who celebrate their Pyrrhic victory on the ashes of past and present vegetation and other forms of life, which they commit to fire.

Seeing that, for Aristotle (who still maintains a strong hold on *energeia*, a word that he introduced into the philosophical vocabulary), the prototype of matter is *hylē*, wood or the woods, the violent extraction of energy paints a vivid picture of the relation between matter and spirit prevalent in the West. A flaming spirit sets itself to work by destroying its other and triumphs over the wooden matter it incinerates. The price for the energy released in the process of combustion is the reduction of what is burnt to ash.

It is not that plants are exempt from the general combustibility that, for Schelling, defined the very living of life. They engage directly with the energy of the sun and release oxygen, providing the elemental conditions of possibility for the burning of fire. But the vegetal mode of obtaining energy—especially that of the solar variety—is non-extractive and non-destructive; the plant receives its energy by tending, by extending toward the inaccessible other, with which it does not interfere. That is one of the most important vegetal lessons to be learned: how to energize oneself, following the plants, without annihilating the sources of our vitality.

In the meantime, energy extraction tears both living and dead things apart, penetrating into their core, enucleating them. Energy production is a fury of destruction. It does not relent until the atom is split, until it reaches the nucleus and divides the indivisible. Nuclear power, with the atomic energy it unbridles, is the apotheoses of the contemporary energy paradigm. So is hydraulic fracturing, or fracking, that cracks the earth (particularly shale rocks) open by exerting high water pressure on them from below. Environmentally destructive and shockingly shortsighted as these methods of energy production are, they are not surprising, in view of the prevalent conception of energy that involves breaching and laying bare the depth of things (of the atom, of the earth …) and drawing power from this violent exposure.

On the one hand, most approaches to energy presuppose substantial divergence between the inner and the outer, depth and surface. The very language of "storage" and "release" indicates that the energy of everything from galaxies to microbes, from economic systems to psychic life, is contained (held inside and prevented from achieving its full actuality) before it is liberated with more or less force. The encompassing whole is, likewise, seen as a great container, from which no energy ever escapes; that is what, at bottom, the law of the conservation of energy means. Absent the dimension of interiority, one can no longer explain how things work, how they are put to work, activated, or withheld in potentiality. Energy differentials depend, above all, on the difference between the inside and the outside, on the speed and force with which these boundaries are traversed.

Plants, *on the other hand*, do not need to devastate the interiority of another being to procure their energy. They set to work the elements that they neither control nor dominate nor appropriate. Besides water and the minerals, which they draw from the soil, they receive what they need from the sun, processing their solar sustenance on their maximally exposed surfaces, the leaves. (Mind you, plants can deplete the soil, but this happens only with human inference, due to intensive agriculture and the spread of monocultures. By and large, vegetation returns to the earth much of what it takes in the processes of its decomposition.)

Human reliance on solar energy would indicate our willingness to learn from plants and to accept an essentially superficial mode of existence or, at the very least, to integrate it with the dimension of depth. Although current technical capabilities would sustain a nearly total reliance on elemental energy (solar, wind, hydro, etc.), they do not match the prevalent mindset that construes energy in its essence as something destructive-extractive and snatched from the interiority of things. The focus of attention may, in fact, swing to "clean energy" and that is, in and of itself, laudable. But "cleanness" relates primarily to the effects of its utilization, not to the question of *what energy is*. That is why oil, coal, and, especially, natural gas companies can claim that they are transitioning to clean energy, without radically modifying the sources of fuel themselves, let alone how they are procured.

Be it labor or truth, we draw value from the core of the human and destroy the material "shell" in the process. On economic spreadsheets, we are accounted for as human resources, from which work can be extracted in keeping with the demands of capital. Our epistemologies, too, are driven by the desire to reveal the inner core of reality, usually by shattering and discarding "mere" outward appearances that occlude it. Thinking has assumed the shape of mental fracking. Unless we subscribe to the insights of phenomenology, we are dissatisfied with the surface of things, with how they present themselves to us in everyday life: with all their imperfections, incompletions, shadowy spots, and stamps of finitude. We clear the surfaces of the world away as though they were but a pile of debris, hiding what is really pure and desirable—the energetic core. Metaphysically "clean," this process of energy production is exceptionally polluting when it comes to its physical effects.

For us, superficial actuality, the actuality of the superficies, is never actual enough. As we strive to know what things *really are*, we break them down to atomic and subatomic, chemical and molecular levels, and prevent them from *merely being*. Why would the framework of energy production and extraction be different from that of the production and extraction of knowledge? The two would have to change in tandem, if the human impact on the world, as well as on ourselves, is to be mitigated.

4

WITHOUT CLEAN AIR, WE HAVE NOTHING (*WITH LUCE IRIGARAY*)

A response, written in March 2014, to the worsening global smog situation.

Air pollution is worsening in many places around the world. In Shanghai, such is the oppressive smog, covering the city with a toxic cloud, that authorities have had to install gigantic TV screens to broadcast the sunrise. Salt Lake City has such poor air quality that chemicals in the atmosphere not only give it a different hue but leave residents with a foul, metallic taste in their mouths. Closer to home, Paris has experienced some of its worst levels of air pollution in recent days, while in the European Union as a whole, even at permitted concentrations, industrial and traffic-related pollution is harming cardiovascular health.

Is clean air, along with drinkable water, becoming one of the most precious resources on the planet? Or should we reframe the question and challenge the kind of thinking that converts everything, including the very air we breathe, into economically measurable reserves and commodities?

Today we live in a world so complicated that encountering each other as humans has become almost impossible. However, instead of asking what it means to be human, alleged experts in various domains discuss at great length how to establish coexistence among people. No doubt, such an objective is both relevant and urgent, but in their debates, these experts in peace stray far from a solution, getting lost in technical details without considering the universal sharing of life, especially at the elemental level, from which we could start again. Even if it makes skeptics laugh, we have no viable solution other than to experience the universal human condition as that of a living being, standing naked (despite all the technological accoutrements) in the earth's milieu.

After all, with every breath we take, we expose our lungs to the outside world, regardless of the barriers we have erected between the environment and ourselves. The resistance to envisaging our situation in this way is due to a nihilistic preference for certain powers—be they material or spiritual, capitalist or cultural—over life itself. The predominant preference is both suicidal and murderous, even though few people actually intend it to be so negative. How can they recover their taste for life and learn ways of cultivating it, in themselves and with others?

The fact that public parks in cities become crowded as soon as the sun shines proves that people long to breathe in green, open spaces. They do not all know what they are seeking, but they flock there, nevertheless.

In these surroundings, they are generally both peaceful and peaceable. It is rare to see people fighting in a garden. The elemental struggle unfolds first not on an economic or social stage, but over air, essential to life itself. If human beings can breathe and share air, they don't need to struggle with one another. And, consequently, it appears to be a basic crime against humanity to contribute to air pollution.

Regrettably, in our Western tradition, neither materialist nor idealist theoreticians give enough consideration to this basic condition for life. As for politicians, despite proposing curbs on environmental pollution, they have not yet called for it to be made a crime. Wealthy countries are even allowed to pollute, if they pay for it. But is our life worth anything other than money? Are we, then, still living? Or do we only sense what life could be when we are welcomed into green spaces?

The plant world shows us in silence what faithfulness to life consists of. It also helps us to a new beginning, urging us to care for our breath, not only at a vital, but also at a spiritual level. We must, in turn, care for it, opposing any sort of pollution that destroys both our world and that of plants. The interdependence, to which we must pay the closest attention, is the one that exists between ourselves and the vegetal world. Often described as "the lungs of the planet," the woods that cover the earth offer us the gift of breathable air by releasing oxygen. But their capacity to renew the air polluted by industry has long reached its limit. If we lack the air necessary for a healthy life (or, indeed, for any kind of life), it is because we have filled it with toxic chemicals and undercut the ability of plants to regenerate it. As we know, rapid deforestation combined with the massive burning of fossil fuels, which are largely the remnants of past plants, is an explosive recipe for an irreversible disaster.

The fight over resources will lead the entire planet to an abyss unless humans learn to share life, both with each other and with plants. This task is simultaneously ethical and political, because it can be discharged only when each takes it upon her- or himself and only when it is accomplished together with others. The lesson taught by plants is that sharing life augments and enhances the sphere of the living, while dividing life into so-called natural or human resources diminishes it. We must come to view the air, the plants, and ourselves as contributors to the preservation of life and growth, rather than a mesh of quantifiable objects or productive potentialities at our disposal. It is then that we would finally begin to live, rather than being concerned with bare survival.

5

POLAND'S BIALOWIEZA: LOSING THE FOREST AND THE TREES

In 2017, Poland authorized logging in one of Europe's last primeval forests.

As of January 1 of this year, Poland has become the European frontline in the war on the environment, and, particularly, on plant life. On that day, Szyszko's law, named after Polish Minister of Environment Jan Szyszko, came into effect. According to the provisions of the law, owners need not inform the authorities if they plan to cut down trees growing on their privately held lands, nor are they required to do any replanting to offset tree felling.

The new provisions have been contested both by the European Union and by diverse groups within Polish civil society. Most emblematically, a group of women under the leadership of Cecylia Malik took the initiative to create the "Mothers on Tree Stumps" movement, whose members breastfeed their children while sitting on recently felled trees. The photographs of their collective actions are posted on social media and show a near-apocalyptic atmosphere with a stark contrast between infants, representing hope for the future, and the scenes of devastation around them.

Yet, it seems that Szyszko's law has not satisfied the forestry industry in Poland. Currently, the state itself is engaged in a massive logging program at the UNESCO-protected Bialowieza ecosystem, considered to be Europe's last primeval forest. Nearly 10,000 acres of Bialowieza are earmarked for logging. Here, too, protests have abounded; earlier this month, hundreds of writers and artists sent an open letter to the Polish authorities imploring them to stop the assault on Bialowieza.

What jumps out at us as we take a closer look at official explanations of the government's decision to open a primeval forest "for business" is how they reproduce the self-contradictions apparent in discussions of security and its relation to basic rights. In the name of security, citizens in Western democracies are asked to give up their privacy but, upon doing so, become less, rather than more, secure. Similarly, in the name of the health of the forest, Polish people are urged to let go of that very forest.

Last year, Jadwiga Wisniewska, a member of the European Parliament from the ruling Law and Justice (PiS) party, stated, "We want to save the forest from its false defenders, who would allow the forest to rot away before our eyes."[1] How is it that the forest is saved through a massive logging program? According to the Polish government, these drastic measures are needed in order to stem out a bark beetle outbreak and to "ensure safety" for the roughly 120,000 tourists visiting the forest annually.

Let us put aside for a moment the environmental objections to a plan that is likely to invite more severe beetle outbreaks in whatever remains of Bialowieza and that ignores the capacities of a 9,000-year-old forest to fight infestations and to regenerate itself. Does the rhetoric of those in power in Poland not reiterate the long-standing mantra of the sacrifices to be accepted for our own good? "To protect the forest, we need to chop it down" reiterates the message that, "To protect our democracy, it is imperative to give up basic democratic rights." Political management reverberates, as though in an echo chamber, with forest management.

The stress on public safety is only one of the elements of complex analogies between the control and administration of human and natural resources, well beyond the scope of the economic sphere where these terms tend to circulate. The Polish government, like Donald Trump in the United States, is spearheading the politics of post-truth, which, in the two cases (and despite undeniable differences in national contexts and fine-grained details) boils down to one and the same thing. In their discourse and practices alike, both regimes have given up on minimal consistency and common sense, previously spun by ideological means. Post-truth is, at its core, a post-ideology of barefaced lies that needn't bother hide the collusion and, at times, the merging of political and economic powers. Just as Trump's conflicts of interest are rife, so some have accused Szyszko—a former park warden and professor of forestry—of being "too close to the logging industry."

When Wisniewska spoke out against "allowing the forest to rot away," she ignored the cycles of growth and decay, crucial to forest regeneration and sustainability. Even if those trees that are affected by the bark beetle outbreak are now deadwood, their rotting away is essential to the future life of the forest. As early American environmentalist Aldo Leopold was fond of emphasizing, what might appear useless to us is absolutely non-negotiable for an ecosystem, or, in his words, "for the health of the land."

Far from a personal oversight, Wisniewska's statement encapsulates the position of the Polish government. Szyszko himself has noted that "more than 4 million cubic meters of wood are rotting, depreciating, species are dying, habitats are lost. Meanwhile, the local population has nothing to heat their homes with, and must import dirty coal from Belarus." Nonetheless, vegetal rot is not depreciation but appreciation, the enrichment of the soil, welcoming the growth to come. With environmental logic out of the picture, we are left with a choice that is no choice, one where both alternatives are unacceptable: to burn "dirty coal from Belarus" or the presumably beetle-infested Polish wood; petrified plants that died millions of years ago or those that perished yesterday, due to an infestation or a buzzing chainsaw.

When it comes to energy and plant life, the past ideological epoch and the current era of post-truth are in agreement insofar as they are ready to sacrifice the long-term wellbeing of ecosystems and the entire planet for short-term pragmatic concerns. The disjunction between the short and the long terms shows that we do not (yet) know what is for our own good, let alone who or what this "we" is in the interweaving of human existence with the living fabric of the environment. A similar opacity haunts us as we face the equally fake choice between security and fundamental rights. Opting for either possibility, we lose not the forest for the trees but the forest *and* the trees. The Polish case is only an extreme literalization of this dilemma.

6

JUST RANDOMNESS?

A 2018 essay on the problems of algorithmic justice and its imbrications with randomness.

In 1979, by an overwhelming majority decision grounded in the Fourth Amendment, the US Supreme Court ruled that police had no right to stop drivers at random. These stops, it argued, "violate the constitutional guarantee against unreasonable seizures."[1] Today, almost forty years after the Supreme Court ruling, the Transportation Security Administration (TSA) "randomly" selects some passengers for additional security screening, such as a pat-down by an officer. But what has changed in the intervening period? Why are random traffic stops by police largely impermissible, whereas random detailed searches at airports are authorized and even explicitly written into the rulebook?

The answer is glaringly obvious: unlike drivers stopped by the police, passengers identified for additional airport screening are picked by a computer program, which creates the illusion of control for the human factor operative in police activities. According to this line of thought, the problem is that even apparently haphazard human choices are not random enough, and therefore not disinterested or impartial enough. Algorithms are implicitly and rather automatically (might we say *algorithmically*?) interpreted as corrective measures here. Ostensibly impersonal and indifferent to whomever they are applied to, they are deemed to live up to the exigencies of justice by seeming "more random."

In the ancient world, sortition and the casting of dice or lots (procedures grouped under the heading of cleromancy) were in use at some of the most important points in personal and political life. Election by lots was an integral part of the democratic process in ancient Greece—above all, in Athens. In the Hellenic and Hebraic paradigms alike, the randomness of the outcome was

145

seen as an expression of divine will, which could take care of the future much better, more successfully and wisely than humans with their limited knowledge. Chance stood for a higher necessity, inaccessible to our faulty reasoning and dim awareness of causes and their effects. The Roman goddess Justitia, who later became Lady Justice, was depicted blindfolded, not only suggesting freedom from prejudice but also guaranteeing that divine indifference would neutralize biases alongside familial, affective, and other kinds of attachments that inevitably persist in human decision-making.

One can imagine a modern instantiation of sortition in public life: electoral tie-breaks decided by casting lots, for instance, or the randomization of waiting lists for organ donations. More often, however, our hopes of deliverance from bias are transferred onto algorithmic decision-making systems, which have been broadly implemented across contemporary societies in the hopes of making employment, financial, legal, and other decisions fairer. Many managers of human resources, for instance, resort to data-driven algorithms in order to sift through the pools of job candidates and make appropriate hiring decisions. The gods of old have been carried over into the present and the future in the shape of computational thinking, artificial intelligence, and technological innovation. Whereas many critics have pointed out how algorithmic systems often conserve rather than eradicate bias, stubborn faith in their superhuman ability to correct essential flaws in our human condition persists. They allow people to "recuse" themselves from decision-making processes and avoid making sense of causal relationships and phenomena when these are too complex to parse. As a result, human actors believe they have mitigated their biases, as though prejudiced thinking could not be transmitted to and engrained in an automated process.

Excessive reliance on algorithms simultaneously masks the persistence of bias and threatens to make human experience itself appear totally random. The streamlining of algorithmic processes in everyday life can make it seem like the milestones of your existence, such as getting a job (or, more often, receiving a rejection letter in response to a job application), befell you out of the blue, with no rhyme or reason, with no one to blame, to praise, or to hold responsible. Would you like to live in a world where everything happened without a *why* and a *because*? What would life feel like, were you to perceive it, including every major and minor occurrence it was woven of, as part of a strange lottery? How would you string together the story of such a life? What, if anything, would there be to narrate? Where would the descriptors "good" and "bad," "just" and "unjust," belong in this mess? Does justice have any meaning outside of human deliberation?

Despite "randomization" techniques in security screenings at airports, passengers deemed "high-risk" often experience discrimination on the basis of their ethnicities, nationalities, religious affiliations, or names. When all is said and done, the parameters for a computer program are set by human programmers. Algorithmic bias preserves sexist, racist, and other problematic attitudes, and may actually amplify already existing prejudices. AI is also prone to learning to be sexist and racist through repeated interactions with humans espousing such attitudes. Although these structural flaws are widely publicized, there are no signs of deep disappointment with them, because they displace the responsibility for biased conduct onto an impersonal system.

Introducing true randomness into our social, political, and legal realities would dovetail with the dream of modernity to reorganize human institutions on a scientific foundation. If chaos theory applies to the universe at large, then why should it not have any influence on the legal system? If scientific experiments use randomization, why shouldn't experiments in living together do the same? Religious faith, which in the course of the European Enlightenment has mutated into faith in reason, is now imperceptibly passing into faith in technology. Technocratic solutions to the nearly universal problems of human existence and coexistence are the legatees of these ideological mutations.

By attempting to prevent selection bias, randomization techniques in clinical trials (allocation concealment, blinding, etc.) unclutter cause-effect relations in the assessment of experimental treatments. They ensure that (1) the tested subjects in various groups are not "systematically different" and (2) no prior knowledge of group assignment, capable of influencing the outcome, exists. But transferring such procedural protocols to legal and political systems (say, by blinding persecutors to the defendants' race) would not reduce selection bias so much as transmute an actual suspect into a neutral, genderless, raceless, classless, abstract, and context-free individual, befitting a technocratic outlook. Unlike in the life and medical sciences, "blinding"—assuming such a thing were possible—neither respects the diverse fabric of social life nor gives way to the transparency of knowing the cause, if only a probable one, behind a determinate effect. Lady Justice's blindfold stays on when it should have been already removed.

The idea that a "random" system could be just fuels the development of sociopolitical algorithms and the emphasis of quantification that they require. To achieve equal treatment under the law through randomization, each citizen must be first converted into a number, giving equality a strictly mathematical or statistical expression and undermining its substantive dimension. Human relations are missing from the purview of quantitative equality, not only at the level of causes and effects but also at the level of community, reduced to an amorphous pool of individuals, from which aleatory samples are drawn for various social proceedings, as in jury pools.

Instead, randomness enables efficiency to override other concerns. At its source in the Old French *randon*, the word *random* itself means "swiftness," a "great rush," from the verb *randonner*, "to run fast." In terms of efficiency, overviewing everything quantitatively is quicker than going into painstaking details of diverse cases and circumstances. The price we pay for efficiency is exorbitant, though; the bill arrives in the form of blindness to singularity, the blindness that flips justice around into injustice.

And yet, the kind of technologically mediated "randomness" such systems offer also harks back to the difficulties we face in attributing causality. We live in an age when direct attributions of causes and effects are exceptionally difficult, because the former are more and more distant from the latter in space and in time. Randomness may then be perceived as the overarching principle of events experienced as though coming from nowhere, happening as effects of effects untraceable to a determinate cause. Given this state of affairs, reliance on randomization procedures may be more palatable, compared with arguments from necessity that secure, beyond a shadow of doubt, the link between a cause and its effects.

The ancient worldview again proves instructive here. In the Book of Job, the divine punishments that befall Job are not the effects of a cause found in his moral conduct; what he undergoes is a purely random experience. But the cause is not absent (God has his reasons); it is simply beyond human understanding. An analogous sentiment today renders the *prima facie* "randomness" of algorithmic systems more acceptable. Given the opacity of iterative machine learning and the mass of data combed for correlations, it seems that an algorithm's output is as "random" as God appears to be in the Book of Job. The current state of technology asks us to have faith in the God-like algorithms that, for all intents and purposes, operate beyond human understanding. The carryover of past theological beliefs shorn of their original context, then, makes us all too willing to accept the framework of *justice without justifications*.

The ends of justice—the aims and goals it pursues—are sacrificed to the smooth functioning of the means when we introduce randomness into the equation. Without a *why* and a *because* the foundations for justice, for legal and political procedure, or, broadly, for human coexistence are dubious. On the level playing field of equality before the law posited by a purely quantitative approach, a choice appears just when, randomly generated, it eludes narrative justifications.

The ideal of randomness and the algorithmic shaping of social reality undercut the very possibility of ordering and reordering the world, along with the ongoing search for meaning. Whether or not our latest Weltanschauung represents the universe as a chaotic mess, splintered at the extreme into an infinite number of pluriverses, a meaningful life unfolds with a modicum of (a flexible, adjustable or suspendable, mutable) order.

Rather than a correctly or incorrectly implemented set of procedures, justice is entwined with the view of what reality is and of the way things should be in the best of possible scenarios. In pluralistic societies, where views frequently clash, ongoing discussions, deliberations, and sensible explanations for decision-making are indispensable. Bypassing these elements under the cover of randomness and algorithmic solutions serves only to mask—not to resolve—the underlying conflicts of interpretation corresponding to frictions between different ways of life or basic worldviews.

The issue, as I see it, is not the use of inevitably biased algorithms or the ideal of randomness that provides an alibi for them, but the blind worship of algorithms without considering substantive, qualitative, causal, and other concerns. The danger is that, with overreliance on automated decision-making systems, there will be no need for thinking (and perpetually rethinking) through justice either in its general outlines or in its application to particular cases. Computerized calculation will take the place of deliberation and will be responsible for further concealing unjust intentions programmed into the "algorithms of justice." Few things are more harmful to our vital ability to seek out the meanings of the world and of our place in it than that.

7

THE IDEA OF FOLLOWING IN THE AGE OF TWITTER

A 2012 analysis of "following" in the context of social media.

Samuel Beckett's *Waiting for Godot* has become a perfect allegory for the globalized world. Always on our guard, we expect something exceptional to happen at any time of day or night, and the absurd fascination with this shapeless possibility binds us to social networks with all the force of affective attachment Freud termed *cathexis*. We intuitively know that the event will arrive by email, in a text message, or in a tweet, each of them potentially unheard-of, life-changing, radically new.

More often than not, what does arrive is a weekly 10 percent discount from a trendy clothes store or a status update on how utterly bored your friend is. Contemporary communication technologies have much more to do with pure possibilities than with what they actually convey, which is why the gap between what is and what could be communicated fails to shock us. And, because we receive even the least significant messages in a heightened state of expectation, it is difficult to ignore them entirely. Instead, they leave a powerful imprint directly on our unconscious, by now brimming with digital debris and shadowy remainders of hi-tech part objects.

Given the combination of psychic overinvestment with open-ended expectations, the new media are saturated with nearly messianic connotations. Hurriedly written books in the genre of "tweeting-the-revolution" that mushroomed after the events of the Arab Spring testify to this—perhaps necessary—overestimation. But what are the more concrete social and political consequences of Tweeter, Facebook, Instagram, and so forth? How, for instance, are they changing right before our eyes such basic power relations as leading and following?

From every corner, one hears calls: "Follow us on [fill in the blank with your preferred social network]!" (Having banned such reminders from its airways, France is a notable exception here.) The implication of this appeal is that, if you do not follow, you will be out of the loop and at a disadvantage, deprived of access to the valuable commodity that is information. But—truth be told—it is the number of virtual followers an individual or a company boasts that makes for its social capital, not vice versa. The initial order, "Follow!" betrays the tacit dependence of those who issue it on their present and future followers. It is symptomatic of the workings of ideology in the digital age.

With regard to the relativity of value, Marx expressed this function of ideology in the clearest terms in Volume I of *Capital*: "One man is king only because other men stand in the relation of subjects to him. They, on the contrary, imagine that they are subjects because he is king."[1] It is up to us to translate Marx's dialectical insight into a couple of simple formulas, according to which (1) the balance of your influence is positive if you have more followers than the number of people you, yourself, follow and (2) this influence resides not in the one followed, but in the recognition bestowed by the followers. To "unfollow" or to "unfriend" someone is a huge insult, a gesture that breaks the distorted looking glass of ideology and demonstrates the power of the follower over the one followed. No wonder, then, that the media treat celebrities unsubscribing from the feeds of other celebrities as newsworthy events!

Twentieth-century totalitarianisms still relied on the ideological constructions Marx, in the nineteenth century, was familiar with: they entailed top-down chains of command comprising the leading—with the Leader at the highest point in the hierarchy—and the led. By contrast, the commerce between the following and the followed in the twenty-first century paints the image not of a vertical system, but of a de-centered, horizontal, reversible arrangement, presumably conducive to a genuine democracy. While the Leader's power was the origin of the political system, in online networks it is no longer clear where such origins reside. But this is not to say that they have evaporated—only that they have become more thoroughly displaced and hidden.

In the dispersion of the network, even when social capital peaks, amounting to tens of millions of followers, all that remains is the formal and quantitative difference—which is also the objective measure of power—between following and being followed. Branching out in every conceivable direction, the network paints an alluring image of anarchy beneath the veil of blurred socio-political relations. In its dispersion, power appears both to undergo dematerialization and to dissipate, in light of the formal equality of anyone with a Twitter or Facebook account.

Not only the origin but also the end, or the purpose, seems irrelevant to the idea of following in the age of digitally reproducible social and political relations. Traditionally, trailing a master guide helped the apprentices to achieve a particular goal: for instance, to increase their knowledge or to improve their skills. Some of the most emblematic narratives in the West, such as *The Divine Comedy* where Dante and his readers both literally and figuratively follow Virgil through the circles of the Inferno (as well as the Purgatory) and Beatrice through the spheres of Heaven, gave voice to the master-apprentice relation. However lengthy they were, these journeys had an end that corresponded to the accomplishment of particular objectives.

Compare this to following someone or something on Twitter or Facebook. Unlike goal-oriented—and, therefore, terminal—apprenticeships, these relations do not have an inherent end, unless for whatever reason you decide to terminate them, hitting the "unfollow" or the "unfriend" button. In their open-endedness, they imitate life, which has neither a guide nor a final outcome because death is not its culmination, but rather an interruption. More than that, like human living itself, following and learning are not purely passive behaviors. It is necessary to know how to follow others, so as to emancipate oneself from this somewhat subservient relation. And yet, in the digital world, the active component of following is absent, as we are drawn along, more or less haphazardly, by whatever is "trending" at the moment. The more we practically follow others, the less we know how to follow, or what following even means.

It is relatively easy to reconcile a following devoid of distinct leadership and goals with the Western ideology of individualism. Social networks create the illusion of a community free of conformism: you can choose exactly who you wish to follow, just as consumers are able to exercise their right to purchase this or that commodity on the market. The sum total of those you follow is supposed to be the expression of your personality, of your individual tastes, preferences, and styles. These, however, are not exempt from the logic of the market, let alone of marketing, which is why the most massive followings gather around the most commodified figures, namely pop or sports stars.

The existence of followers is enmeshed with the digital lives of those they follow, furnishing evidence of *cathexis* and affective attachment. What counts is the possibility of influence over followers, not this or that instance of imitation. Potentiality is, indeed, the capital of social networks. Facebook stocks have already made their lackluster debut on Nasdaq, where one has a chance to trade in digital potentiality itself. The idea of following in the age of Twitter has now come into its own; it means, invariably, "Follow us on the stock exchange!"

THE UPHEAVALS YET TO COME

The retrospective glance I have cast in these pages on the second decade of the twenty-first century does not overview past world turbulence from the position of relative stability and security in the present. If anything, drastic changes are accelerating on the political, cultural, intellectual, and technological planes. The four kinds of upheaval travel together—akin to the four horsemen of the apocalypse—and feed off one another. They shape our shared reality by disfiguring and reconfiguring it through the dynamics of catastrophic climate change and a pandemic, the global rise of the extreme right and new impositions of "woke" political correctness, genetic manipulation and advances in AI.

The above list and the themes broached in the chapters of this collection are certainly incomplete, but they do bring together some of the most salient tendencies that have destabilized and will keep destabilizing the world for the time to come. It turns out that upheaval is much more than a sudden and traumatic event, disturbing the stagnant waters of history. More than a disruption of incremental history, upheaval has a history of its own. While destabilizing and disorienting all those caught in its crosshairs, upheaval extends into the future, productive not so much of finite time as of time's very finitude.

So, what is our future made of upheavals, their effects amplified by multiple resonances among them, going to look and feel like? Philosophy is not (or, at least, it should not be) in the business of social or planetary chiromancy. The indelible allegory of the Owl of Minerva that, taking flight at dusk, overviews and knows the late fruit of the world is still, to my mind, as suitable for the philosophical endeavor as it was in the times of Hegel, who included this image in the introduction to his *Philosophy of Right*. But a backward-looking gaze is, simultaneously, staring forward, catching sight of future upheavals as extensions of the past and the present. In this Janus-faced capacity, philosophy forms a speculative fold in actuality, the fold of self-understanding through estrangement from a simple and unproblematic division between the individual and collective, human and other-than-human "self."

Seen from the philosophical standpoint, the upheavals yet to come do not refer to future pandemics and social unrest, to new advances in information technologies and to international military or paramilitary conflicts over territories and dwindling resources. To be sure, all these are the highly probable aspects of major upheavals in the twenty-first century. But they are also, and more importantly, the practical expressions of underlying conceptions, of ideational frameworks that go hand in hand with and, in part, steer events in the world.

Future pandemics and catastrophic climate change, for example, are derivative with regard to the ideology that treats natural environment as a reserve to be tapped into at will. Armed with this ideology, human infringements on animal and plant species living in the wild or on the elemental make-up of the earth, the oceans, and the atmosphere drives the upheavals in question. AI is also likely to mushroom, posing ethical quandaries and defining our everyday lives, because our implicit notion of intelligence is, in and of itself, artificial. According to the construal of contemporary cognitive sciences, "natural" human intelligence is already mediated through algorithms, as the latest incarnation of calculative rationality, and made to conform to the informatic paradigm in general. That is why the intelligence called artificial can ooze, seep, or freely flow into our existence (say, with a project such as Neuralink): AI is the exact same thing as the current scientific construction of human thought processes, just externalized and exaggerated.

If I were to recap them in a single phrase, I would say that the upheavals yet to come consist of multiple torsions and tensions of an inherently frustrated possession: the possession of our natural milieu, of national territories, of extraordinary technological capacities, and, last but not least, of ourselves as human beings. Political, cultural, intellectual, and technological grounds are heaving beneath our feet as attempts to capture and dominate the world (or, more accurately, world*s*) backfire, taking hold of *us*, who are not situated at a safe distance from the desired object of appropriation. Future upheavals converge on this paradoxical condition, in which the consolidation of control renders reality more uncontrollable, our grasp upset, achieving the opposite of the intended effect.

NOTES

A SENSE OF UPHEAVAL

1 Karl Marx, "Theses on Feuerbach," in Robert C. Tucker (ed.), *The Marx-Engels Reader*, 2nd ed. (London: W.W. Norton, 1978), p. 145.
2 Michael Marder, *Dump Philosophy: A Phenomenology of Devastation* (New York: Bloomsbury, 2020); Michael Marder, *Pyropolitics in the World Ablaze* (London: Rowman & Littlefield, 2020).

PART I POLITICAL UPHEAVAL

THE UNFINISHED COLLAPSE OF THE SOVIET UNION

1 "Russia's Neighborhood in Flames," *Financial Times*, October 2020. https://www.ft.com/content/84ec2707-e829-4bc6-b81c-e19453883354.
2 Clara Ferreira Marques, "Russia's Post-Soviet Hegemony Is Fading," *Bloomberg News*, October 9, 2020. https://www.bloomberg.com/opinion/articles/2020-10-09/armenia-belarus-kyrgyzstan-russia-s-post-soviet-hegemony-is-fading.
3 Anton Troianovski, "Putin, Long the Sower of Instability, Is Now Surrounded by It," *New York Times*, October 7, 2020. https://www.nytimes.com/2020/10/07/world/europe/putin-belarus-kyrgyzstan-caucasus.html.

INCENDIARY WORDS AND THE VOLCANO OF OCCUPATION

1 Natan Chofshi, "Into the Abyss," in M. Buber, J. L. Magnes, and E. Simon (eds.), *Towards Union in Palestine: Essays on Zionism and Jewish-Arab Cooperation* (Westport: Greenwood Press, 1972), p. 40.

CAN THERE BE POETRY AFTER NETANYAHU?

1 Theodor Adorno, *Prisms* (Cambridge, MA: MIT Press, 1983), p. 34.

THE EUROPEAN UNION AND THE RHETORIC OF IMMATURITY

1 Quoted in K. Polanyi, *The Great Transformation: The Political and Economic Origins of our Time* (Boston, MA: Beacon Press, 1957), p. 113.

TRUMP METAPHYSICS

1 Linda Martin Alcoff, "Why Trump Is Still Here," *The Philosophical Salon*, February 8, 2016. https://thephilosophicalsalon.com/why-trump-is-still-here/.

THE CON ARTISTRY OF THE DEAL: TRUMP, THE REALITY TV PRESIDENT

1 Donald Trump, *The Art of the Deal* (New York: Ballantine Books, 2015), p. 58.
2 William Egginton & David Castillo, *Medialogies: Reading Reality in the Age of Inflationary Media* (New York: Bloomsbury, 2016).
3 Trump, *The Art of the Deal*, p. 60.
4 Slavoj Žižek, "To Paraphrase Stalin: They Are Both Worse." *In These Times*, November 6, 2016, https://inthesetimes.com/features/zizek_clinton_trump_lesser_evil.html.
5 Oliver Milman, "Donald Trump Presidency a 'Disaster for the Planet,' Warn Climate Scientists," *Guardian*, November 11, 2016. https://www.theguardian.com/environment/2016/nov/11/trump-presidency-a-disaster-for-the-planet-climate-change.

CAN DEMOCRACY SAVE THE PLANET?

1 Jeffrey Sachs, "America's Broken Democracy," *Project Syndicate*, May 31, 2017. https://www.project-syndicate.org/commentary/trump-political-masters-koch-brothers-by-jeffrey-d-sachs-2017-05.
2 Private conversation.

PART II CULTURAL UPHEAVAL

THE TWO SUNS OF EUROPE

1 G. W. F. Hegel, *Philosophy of Nature: Encyclopedia of the Philosophical Sciences, Part II*, translated by A. V. Miller (Oxford: Oxford University Press, 2004), 306.

PART III INTELLECTUAL UPHEAVAL

1 Patricia Cohen, "An Ethical Question: Does a Nazi Deserve a Place among the Philosophers?" *New York Times*, November 8, 2009. https://www.nytimes.com/2009/11/09/books/09philosophy.html.

2 Michael Marder & Marcia Sá Cavalcante Schuback, "Philosophy without Right? Some Notes on Heidegger's Notes for the 1934–5 'Hegel Seminar'," in Peter Trawny, Marcia Sá Cavalcante Schuback, and Michael Marder (eds.), *On Hegel's* Philosophy of Right: *The 1934–35 Seminar and Interpretative Essays* (London: Bloomsbury, 2014), p. 84.

HEIDEGGER'S THINKING TODAY IS, PERHAPS, THE POSSIBILITY OF THE WORLD

1 Martin Heidegger, *Bremer und Freiburger Vorträge*, 2nd ed. (Frankfurt-am-Main: Vittorio Klostermann, 2005), p. 88.

PLUS DE RESTES: REMEMBERING JACQUES DERRIDA

1 Michael Behrent and Héloïse Lhérété, "*Que rest-t-il de Jacques Derrida?*" *Sciences Humaines*, September 15, 2014. http://www.scienceshumaines.com/que-reste-t-il-de-jacques-derrida_fr_33151.html.
2 Benoît Peeters, *Derrida: A Biography* (Cambridge, UK: Polity Press, 2013), p. 541.
3 Jacques Derrida, *The Work of Mourning*, translated by Pascale-Anne Brault and Michael Naas (Chicago: University of Chicago Press, 2001), p. 217.
4 Jacques Derrida, "Biodegradables: Seven Diary Fragments," translated by Peggy Kamuf, *Critical Inquiry 15*, Summer 1989, p. 824.

NATURALIZE THIS! ANALYTIC PHILOSOPHY AND THE LOGIC OF REACTIVE NEUTRALIZATION

1 I consider this critical bent in detail in my *Phenomena—Critique—Logos: The Project of Critical Phenomenology* (London: Rowman & Littlefield International, 2014).

POSITION AS A POLITICAL CATEGORY: PHENOMENOLOGY AND THE EROTICISM OF POWER

1 Michael Marder, *Political Categories: Thinking Beyond Concepts* (New York: Columbia University Press, 2019).

THE POWERLESSNESS OF PHILOSOPHY

1 Pierre Hadot, "The Ancient Philosophers," in *The Selected Writings of Pierre Hadot: Philosophy as Practice* (London: Bloomsbury, 2020), p. 50.
2 Hadot, "The Ancient Philosophers," p. 51.
3 Hadot, "The Ancient Philosophers," p. 51.

PART IV TECHNOLOGICAL UPHEAVAL

POLAND'S BIALOWIEZA: LOSING THE FOREST AND THE TREES

1 Aleksandra Eriksson, "Poland Seeks Support for Logging in Ancient Forest," *EU Observer*, September 1, 2016. -https://euobserver.com/environment/134841. All subsequent quotes in the present chapter are drawn from this document.

JUST RANDOMNESS?

1 Andrew Glass, "SCOTUS Rules Police May Not Stop Motorists at Random without Cause, March 27, 1979," *Politico*, March 27, 2016. https://www.politico.com/story/2016/03/this-day-in-politics-march-27-1979-221209.

THE IDEA OF FOLLOWING IN THE AGE OF TWITTER

1 Karl Marx, *Capital*, vol. 1 (London: Penguin, 1974), p. 63.

www.ingramcontent.com/pod-product-compliance
Lightning Source LLC
Chambersburg PA
CBHW021143230426
43667CB00005B/240